BRITANNICA

Mathematics in Context

W9-DJP-813

Triangles and Beyond

Britannica

ENCYCLOPÆDIA BRITANNICA EDUCATIONAL CORPORATION

Mathematics in Context is a comprehensive curriculum for the middle grades. It was developed in collaboration with the Wisconsin Center for Education Research, School of Education, University of Wisconsin–Madison and the Freudenthal Institute at the University of Utrecht, The Netherlands, with the support of National Science Foundation Grant No. 9054928.

 National Science Foundation

Opinions expressed are those of the authors
and not necessarily those of the Foundation

ISBN 0-7826-1535-X
1 2 3 4 5 WK 02 01 00 99 98

The *Mathematics in Context* Development Team

Mathematics in Context is a comprehensive curriculum for the middle grades. The National Science Foundation funded the National Center for Research in Mathematical Sciences Education at the University of Wisconsin–Madison to develop and field-test the materials from 1991 through 1996. The Freudenthal Institute at the University of Utrecht in The Netherlands, as a subcontractor, collaborated with the University of Wisconsin–Madison on the development of the curriculum.

The initial version of *Triangles and Beyond* was developed by Anton Roodhardt and Jan Auke de Jong. It was adapted for use in American schools by Laura J. Brinker, James A. Middleton, and Aaron N. Simon.

National Center for Research in Mathematical Sciences Education Staff

Thomas A. Romberg
Director

Joan Daniels Pedro
Assistant to the Director

Gail Burrill
Coordinator
Field Test Materials

Margaret R. Meyer
Coordinator
Pilot Test Materials

Mary Ann Fix
Editorial Coordinator

Sherian Foster
Editorial Coordinator

James A. Middleton
Pilot Test Coordinator

Margaret A. Pligge
First Edition Coordinator

Project Staff

Jonathan Brendefur
Laura J. Brinker
James Browne
Jack Burrill
Rose Byrd
Peter Christiansen
Barbara Clarke
Doug Clarke
Beth R. Cole

Fae Dremock
Jasmina Milinkovic
Kay Schultz
Mary C. Shafer
Julia A. Shew
Aaron N. Simon
Marvin Smith
Stephanie Z. Smith
Mary S. Spence
Kathleen A. Steele

Freudenthal Institute Staff

Jan de Lange
Director

Els Feijs
Coordinator

Martin van Reeuwijk
Coordinator

Project Staff

Mieke Abels
Nina Boswinkel
Frans van Galen
Koeno Gravemeijer
Marja van den Heuvel-Panhuizen
Jan Auke de Jong
Vincent Jonker
Ronald Keijzer

Martin Kindt
Jansie Niehaus
Nanda Querelle
Anton Roodhardt
Leen Streefland
Adri Treffers
Monica Wijers
Astrid de Wild

Acknowledgments

Several school districts used and evaluated one or more versions of the materials: Ames Community School District, Ames, Iowa; Parkway School District, Chesterfield, Missouri; Stoughton Area School District, Stoughton, Wisconsin; Madison Metropolitan School District, Madison, Wisconsin; Milwaukee Public Schools, Milwaukee, Wisconsin; and Dodgeville School District, Dodgeville, Wisconsin. Two sites were involved in staff developments as well as formative evaluation of materials: Culver City, California, and Memphis, Tennessee. Two sites were developed through partnership with Encyclopædia Britannica Educational Corporation: Miami, Florida, and Puerto Rico. University Partnerships were developed with mathematics educators who worked with preservice teachers to familiarize them with the curriculum and to obtain their advice on the curriculum materials. The materials were also used at several other schools throughout the United States.

We at Encyclopædia Britannica Educational Corporation extend our thanks to all who had a part in making this program a success. Some of the participants instrumental in the program's development are as follows:

Allapattah Middle School
Miami, Florida
Nemtalla (Nikolai) Barakat

Ames Middle School
Ames, Iowa
Kathleen Coe
Judd Freeman
Gary W. Schnieder
Ronald H. Stromen
Lyn Terrill

Bellerive Elementary
Creve Coeur, Missouri
Judy Hetterscheidt
Donna Lohman
Gary Alan Nunn
Jakke Tchang

Brookline Public Schools
Brookline, Massachusetts
Rhonda K. Weinstein
Deborah Winkler

Cass Middle School
Milwaukee, Wisconsin
Tami Molenda
Kyle F. Witty

Central Middle School
Waukesha, Wisconsin
Nancy Reese

Craigmont Middle School
Memphis, Tennessee
Sharon G. Ritz
Mardest K. VanHooks

Crestwood Elementary
Madison, Wisconsin
Diane Hein
John Kalson

Culver City Middle School
Culver City, California
Marilyn Culbertson
Joel Evans
Joy Ellen Kitzmiller
Patricia R. O'Connor
Myrna Ann Perks, Ph.D.
David H. Sanchez
John Tobias
Kelley Wilcox

Cutler Ridge Middle School
Miami, Florida
Lorraine A. Valladares

Dodgeville Middle School
Dodgeville, Wisconsin
Jacqueline A. Kamps
Carol Wolf

Edwards Elementary
Ames, Iowa
Diana Schmidt

Fox Prairie Elementary
Stoughton, Wisconsin
Tony Hjelle

Grahamwood Elementary
Memphis, Tennessee
M. Lynn McGoff
Alberta Sullivan

Henry M. Flagler Elementary
Miami, Florida
Frances R. Harmon

Horning Middle School
Waukesha, Wisconsin
Connie J. Marose
Thomas F. Clark

Huegel Elementary
Madison, Wisconsin
Nancy Brill
Teri Hedges
Carol Murphy

Hutchison Middle School
Memphis, Tennessee
Maria M. Burke
Vicki Fisher
Nancy D. Robinson

Idlewild Elementary
Memphis, Tennessee
Linda Eller

Jefferson Elementary
Santa Ana, California
Lydia Romero-Cruz

Jefferson Middle School
Madison, Wisconsin
Jane A. Beebe
Catherine Buege
Linda Grimmer
John Grueneberg
Nancy Howard
Annette Porter
Stephen H. Sprague
Dan Takkunen
Michael J. Vena

Jesus Sanabria Cruz School
Yabucoa, Puerto Rico
Andreíta Santiago Serrano

John Muir Elementary School
Madison, Wisconsin
Julie D'Onofrio
Jane M. Allen-Jauch
Kent Wells

Kegonsa Elementary
Stoughton, Wisconsin
Mary Buchholz
Louisa Havlik
Joan Olsen
Dominic Weisse

Linwood Howe Elementary
Culver City, California
Sandra Checel
Ellen Thireos

Mitchell Elementary
Ames, Iowa
Henry Gray
Matt Ludwig

New School of Northern Virginia
Fairfax, Virginia
Denise Jones

Northwood Elementary
Ames, Iowa
Eleanor M. Thomas

Orchard Ridge Elementary
Madison, Wisconsin
Mary Paquette
Carrie Valentine

Parkway West Middle School
Chesterfield, Missouri
Elissa Aiken
Ann Brenner
Gail R. Smith

Ridgeway Elementary
Ridgeway, Wisconsin
Lois Powell
Florence M. Wasley

Roosevelt Elementary
Ames, Iowa
Linda A. Carver

Roosevelt Middle
Milwaukee, Wisconsin
Sandra Simmons

Ross Elementary
Creve Coeur, Missouri
Annette Isselhard
Sheldon B. Korklan
Victoria Linn
Kathy Stamer

St. Joseph's School
Dodgeville, Wisconsin
Rita Van Dyck
Sharon Wimer

St. Maarten Academy
St. Peters, St. Maarten, NA
Shareed Hussain

Sarah Scott Middle School
Milwaukee, Wisconsin
Kevin Haddon

Sawyer Elementary
Ames, Iowa
Karen Bush Hoiberg

Sennett Middle School
Madison, Wisconsin
Brenda Abitz
Lois Bell
Shawn M. Jacobs

Sholes Middle School
Milwaukee, Wisconsin
Chris Gardner
Ken Haddon

Stephens Elementary
Madison, Wisconsin
Katherine Hogan
Shirley M. Steinbach
Kathleen H. Vegter

Stoughton Middle School
Stoughton, Wisconsin
Sally Bertelson
Polly Goepfert
Jacqueline M. Harris
Penny Vodak

Toki Middle School
Madison, Wisconsin
Gail J. Anderson
Vicky Grice
Mary M. Ihlenfeldt
Steve Jernegan
Jim Leidel
Theresa Loehr
Maryann Stephenson
Barbara Takkunen
Carol Welsch

Trowbridge Elementary
Milwaukee, Wisconsin
Jacqueline A. Nowak

W. R. Thomas Middle School
Miami, Florida
Michael Paloger

Wooddale Elementary Middle School
Memphis, Tennessee
Velma Quinn Hodges
Jacqueline Marie Hunt

Yahara Elementary
Stoughton, Wisconsin
Mary Bennett
Kevin Wright

Site Coordinators

Mary L. Delagardelle—Ames Community Schools, Ames, Iowa

Dr. Hector Hirigoyen—Miami, Florida

Audrey Jackson—Parkway School District, Chesterfield, Missouri

Jorge M. López—Puerto Rico

Susan Militello—Memphis, Tennessee

Carol Pudlin—Culver City, California

Reviewers and Consultants

Michael N. Bleicher
Professor of Mathematics
University of Wisconsin–Madison
Madison, WI

Diane J. Briars
Mathematics Specialist
Pittsburgh Public Schools
Pittsburgh, PA

Donald Chambers
Director of Dissemination
University of Wisconsin–Madison
Madison, WI

Don W. Collins
Assistant Professor of Mathematics Education
Western Kentucky University
Bowling Green, KY

Joan Elder
Mathematics Consultant
Los Angeles Unified School District
Los Angeles, CA

Elizabeth Fennema
Professor of Curriculum and Instruction
University of Wisconsin–Madison
Madison, WI

Nancy N. Gates
University of Memphis
Memphis, TN

Jane Donnelly Gawronski
Superintendent
Escondido Union High School
Escondido, CA

M. Elizabeth Graue
Assistant Professor of Curriculum and Instruction
University of Wisconsin–Madison
Madison, WI

Jodean E. Grunow
Consultant
Wisconsin Department of Public Instruction
Madison, WI

John G. Harvey
Professor of Mathematics and Curriculum & Instruction
University of Wisconsin–Madison
Madison, WI

Simon Hellerstein
Professor of Mathematics
University of Wisconsin–Madison
Madison, WI

Elaine J. Hutchinson
Senior Lecturer
University of Wisconsin–Stevens Point
Stevens Point, WI

Richard A. Johnson
Professor of Statistics
University of Wisconsin–Madison
Madison, WI

James J. Kaput
Professor of Mathematics
University of Massachusetts–Dartmouth
Dartmouth, MA

Richard Lehrer
Professor of Educational Psychology
University of Wisconsin–Madison
Madison, WI

Richard Lesh
Professor of Mathematics
University of Massachusetts–Dartmouth
Dartmouth, MA

Mary M. Lindquist
Callaway Professor of Mathematics Education
Columbus College
Columbus, GA

Baudilio (Bob) Mora
Coordinator of Mathematics & Instructional Technology
Carrollton-Farmers Branch
Independent School District
Carrollton, TX

Paul Trafton
Professor of Mathematics
University of Northern Iowa
Cedar Falls, IA

Norman L. Webb
Research Scientist
University of Wisconsin–Madison
Madison, WI

Paul H. Williams
Professor of Plant Pathology
University of Wisconsin–Madison
Madison, WI

Linda Dager Wilson
Assistant Professor
University of Delaware
Newark, DE

Robert L. Wilson
Professor of Mathematics
University of Wisconsin–Madison
Madison, WI

Dear Teacher,

Welcome! *Mathematics in Context* is designed to reflect the National Council of Teachers of Mathematics Standards for School Mathematics and to ground mathematical content in a variety of real-world contexts. Rather than relying on you to explain and demonstrate generalized definitions, rules, or algorithms, students investigate questions directly related to a particular context and construct mathematical understanding and meaning from that context.

The curriculum encompasses 10 units per grade level. *Triangles and Beyond* is designed to be the sixth unit in the geometry strand, but the unit also lends itself to independent use—to introduce students to the geometric properties of triangles and other polygons, such as parallelograms.

In addition to the Teacher Guide and Student Books, *Mathematics in Context* offers the following components that will inform and support your teaching:

- *Teacher Resource and Implementation Guide,* which provides an overview of the complete system, including program implementation, philosophy, and rationale

- *Number Tools,* Volumes 1 and 2, which are a series of blackline masters that serve as review sheets or practice pages involving number issues/basic skills

- *News in Numbers,* which is a set of additional activities that can be inserted between or within other units; it includes a number of measurement problems that require estimation.

Thank you for choosing *Mathematics in Context.* We wish you success and inspiration!

Sincerely,

The Mathematics in Context Development Team

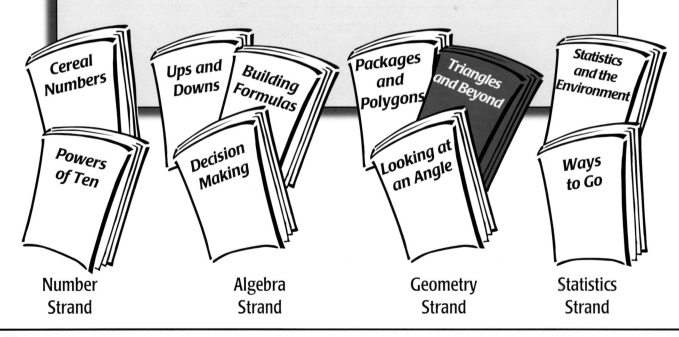

Number Strand

Algebra Strand

Geometry Strand

Statistics Strand

Overview

BRITANNICA

Mathematics
in
Context

How to Use This Book

This unit is one of 40 for the middle grades. Each unit can be used independently; however, the 40 units are designed to make up a complete, connected curriculum (10 units per grade level). There is a Student Book and a Teacher Guide for each unit.

Each Teacher Guide comprises elements that assist the teacher in the presentation of concepts and in understanding the general direction of the unit and the program as a whole. Becoming familiar with this structure will make using the units easier.

Each Teacher Guide consists of six basic parts:

- Overview
- Student Materials and Teaching Notes
- Assessment Activities and Solutions
- Glossary
- Blackline Masters
- Try This! Solutions

Overview

Before beginning this unit, read the Overview in order to understand the purpose of the unit and to develop strategies for facilitating instruction. The Overview provides helpful information about the unit's focus, pacing, goals, and assessment, as well as explanations about how the unit fits with the rest of the *Mathematics in Context* curriculum.

Student Materials and Teaching Notes

This Teacher Guide contains all of the student pages (except the Try This! activities), each of which faces a page of solutions, samples of students' work, and hints and comments about how to facilitate instruction. Note: Solutions for the Try This! activities can be found at the back of this Teacher Guide.

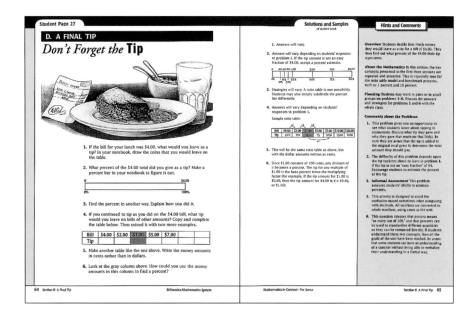

Each section within the unit begins with a two-page spread that describes the work students do, the goals of the section, new vocabulary, and materials needed, as well as providing information about the mathematics in the section and ideas for pacing, planning instruction, homework, and assessment.

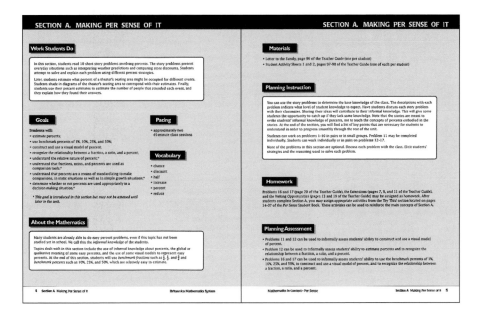

Assessment Activities and Solutions

Information about assessment can be found in several places in this Teacher Guide. General information about assessment is given in the Overview; informal assessment opportunities are identified on the teacher pages that face each student page; and the Assessment Activities section of this guide provides formal assessment opportunities.

Glossary

The Glossary defines all vocabulary words listed on the Section Opener pages. It includes the mathematical terms that may be new to students, as well as words associated with the contexts introduced in the unit. (Note: The Student Book does not have a glossary. This allows students to construct their own definitions, based on their personal experiences with the unit activities.)

Blackline Masters

At the back of this Teacher Guide are blackline masters for photocopying. The blackline masters include a letter to families (to be sent home with students before beginning the unit), several student activity sheets, and assessment masters.

Try This! Solutions

Also included in the back of this Teacher Guide are the solutions to several Try This! activities—one related to each section of the unit—that can be used to reinforce the unit's main concepts. The Try This! activities are located in the back of the Student Book.

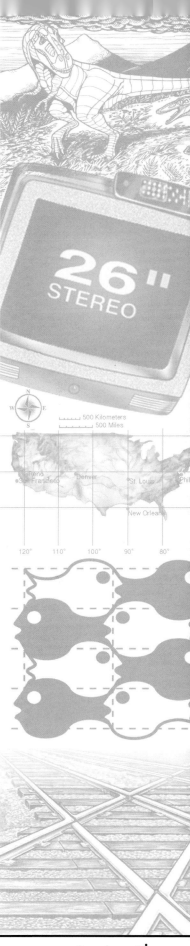

Unit Focus

A key part of the geometry strand is the emphasis on the properties of geometric figures. In this unit, the focus is on triangles.

Students learn to recognize triangles in many different contexts and learn to name different types of triangles. They construct or attempt to construct triangles with various side lengths and formulate a rule about the relationship among the side lengths of a triangle. Another rule students develop in this unit involves the sum of the angles in a triangle. Students also identify the relationship between the side lengths and the angle sizes opposite them. Students learn how to construct triangles and parallel lines, and they identify properties of parallelograms. A final topic introduced in this unit is transformations: Students learn to identify and carry out translations, rotations, and reflections of figures. They rotate triangles to make polygons and parallelograms.

Mathematical Content

- recognizing and using triangles
- describing parallel lines
- describing parallelograms
- identifying translations, rotations, and reflections
- using constructions
- creating visualizations
- identifying congruent figures

Prior Knowledge

This unit assumes that students have:

- identified a triangle,
- measured lengths,
- measured angles with a protractor or compass card,
- learned that there are 180° in a semicircle or straight angle,
- become familiar with properties of regular polygons.

In addition, it will be helpful if students can use a compass to draw a circle.

Planning and Preparation

Pacing: 17 days

Section	Work Students Do	Pacing*	Materials
A. Triangles Everywhere	■ recognize triangles in shapes and in the world	3 days	■ Letter to Family (one per student) ■ Student Activity Sheets 1, 2, and 3 (one of each per student) ■ See page 5 of the Teacher Guide for a complete list of the materials and the quantities needed.
B. The Sides	■ use the relationship between the side lengths of a triangle	3 days	■ Student Activity Sheets 4 and 5 (one of each per student) ■ See page 19 of the Teacher Guide for a complete list of the materials and the quantities needed.
C. The Angles	■ identify the sum of the angles of a triangle as being 180°	3 days	■ See page 37 of the Teacher Guide for a complete list of the materials and the quantities needed.
D. Sides and Angles	■ identify the relationship between the side lengths of a triangle and their opposite angles	2 days	■ Student Activity Sheet 6 (one per student) ■ See page 51 of the Teacher Guide for a complete list of the materials and the quantities needed.
E. Parallel	■ identify properties of parallel lines and parallelograms	2 days	■ Student Activity Sheet 7 (one per student) ■ See page 65 of the Teacher Guide for a complete list of the materials and the quantities needed.
F. Copies	■ identify translations, rotations, and reflections	2 days	■ See page 83 of the Teacher Guide for a complete list of the materials and the quantities needed.
G. Rotating Triangles	■ rotate triangles to make polygons and parallelograms	2 days	■ Student Activity Sheet 8 (one per student) ■ See page 97 of the Teacher Guide for a complete list of the materials and the quantities needed.

* One day is approximately equivalent to one 45-minute class session.

Preparation

In the *Teacher Resource and Implementation Guide* is an extensive description of the philosophy underlying both the content and the pedagogy of the *Mathematics in Context* curriculum. Suggestions for preparation are also given in the Hints and Comments columns of this Teacher Guide. You may want to consider the following:

• Work through the unit before teaching it. If possible, take on the role of the student and discuss your strategies with other teachers.

• Use the overhead projector for student demonstrations, particularly with overhead transparencies of the student activity sheets and any manipulatives used in the unit.

• Invite students to use drawings and examples to illustrate and clarify their answers.

• Allow students to work at different levels of sophistication. Some students may need concrete materials, while others can work at a more abstract level.

• Provide opportunities and support for students to share their strategies, which often differ. This allows students to take part in class discussions and introduces them to alternative ways to think about the mathematics in the unit.

• In some cases, it may be necessary to read the problems to students or to pair students to facilitate their understanding of the printed materials.

• A list of the materials needed for this unit is in the chart on page xiii.

• Try to follow the recommended pacing chart on page xiii. You can easily spend much more time on this unit than the number of class periods indicated. Bear in mind, however, that many of the topics introduced in this unit will be revisited and covered more thoroughly in other *Mathematics in Context* units.

Resources

For Teachers	For Students
Books and Magazines • *Mathematics Assessment: Myths, Models, Good Questions, and Practical Suggestions*, edited by Jean Kerr Stenmark (Reston, Virginia: The National Council of Teachers of Mathematics, Inc., 1991)	**Software** • *The Geometer's Sketchpad*® (available from Key Curricular Press) **Manipulatives** • Miras™ or mirrors for reflections

Assessment

Planning Assessment

In keeping with the NCTM Assessment Standards, valid assessment should be based on evidence drawn from several sources. (See the full discussion of assessment philosophies in the *Teacher Resource and Implementation Guide*.) An assessment plan for this unit may draw from the following sources:

• Observations—look, listen, and record observable behavior.

• Interactive Responses—in a teacher-facilitated situation, note how students respond, clarify, revise, and extend their thinking.

• Products—look for the quality of thought evident in student projects, test answers, worksheet solutions, or writings.

These categories are not meant to be mutually exclusive. In fact, observation is a key part in assessing interactive responses and also a key to understanding the end results of projects and writings.

Ongoing Assessment Opportunities

• Problems within Sections

To evaluate ongoing progress, *Mathematics in Context* identifies informal assessment opportunities and the goals that these particular problems assess throughout the Teacher Guide. There are also indications as to what you might expect from your students.

• Section Summary Questions

The summary questions at the end of each section are vehicles for informal assessment (see Teacher Guide pages 16, 34, 48, 62, 80, 94, and 104).

End-of-Unit Assessment Opportunities

In the back of this Teacher Guide, there are five assessment activities that can be completed in two 45-minute class sessions. For a more detailed description of the assessment activity, see the Assessment Overview (Teacher Guide pages 106 and 107).

You may also wish to design your own culminating project or let students create one that will tell you what they consider important in the unit. For more assessment ideas, refer to the chart on pages xvi and xvii.

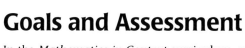

Goals and Assessment

In the *Mathematics in Context* curriculum, unit goals, categorized according to cognitive procedures, relate to the strand goals and to the NCTM Curriculum and Evaluation Standards. Additional information about these goals is found in the *Teacher Resource and Implementation Guide.* The *Mathematics in Context* curriculum is designed to help students develop their abilities so that they can perform with understanding in each of the categories listed below. It is important to note that the attainment of goals in one category is not a prerequisite to attaining those in another category. In fact, students should progress simultaneously toward several goals in different categories.

	Goal	Ongoing Assessment Opportunities		End-of-Unit Assessment Opportunities
Conceptual and Procedural Knowledge	**1.** recognize triangles in the world	**Section A**	p. 8, #2	Do You See the Mathematics?, p. 138
	2. use properties of triangles (such as the sum of angles, side relationships, and the Hinge theorem)	**Section B** **Section C** **Section D**	p. 30, #12 p. 34, #15 p. 48, #13, #14 p. 56, #5, p. 62, #11	The Dragline, p. 139
	3. construct a triangle with given side lengths	**Section B**	p. 30, #12, p. 34, #14	The Rolling Triangle, pp. 133–134
	4. use construction when appropriate	**Section B** **Section E**	p. 34, #15 p. 70, #4	The Dragline, p. 139
	5. understand and identify line symmetry	**Section F**	p. 92, #9 p. 94, #15	
	6. identify congruent figures	**Section G**	p. 104, #8b	

	Goal	Ongoing Assessment Opportunities	End-of-Unit Assessment Opportunities

Reasoning, Communicating, Thinking, and Making Connections	**7.** describe geometric figures using words and/or diagrams	**Section B** p. 24, #5
		             p. 34, #14
		Section D p. 60, #9
		Section E p. 80, #15, #16
		Section G p. 102, #5
		             p. 104, #8a
			The Rolling Triangle, pp. 133–134
			Copies, pp. 135–136
			Rhombus, p. 137
			Do You See the Mathematics?, p. 138

	8. describe transformations (translation, rotation, and reflection) using words and/or diagrams	**Section F** p. 88, #5 p. 94, #14	The Rolling Triangle, pp. 133–134 Copies, pp. 135–136 Do You See the Mathematics?, p. 138

	Goal	Ongoing Assessment Opportunities	End-of-Unit Assessment Opportunities
Modeling, Nonroutine Problem-Solving, Critically Analyzing, and Generalizing	**9.** appreciate that geometry is a means of describing the world	**Section A** p. 8, #2	Do You See the Mathematics?, p. 138 The Dragline, p. 139
	10. create mathematical representations from visualization	**Section D** p. 56, #5	The Rolling Triangle, pp. 133–134 Copies, pp. 135–136
	11. recognize the need for mathematical rigor in justifying answers	**Section E** p. 70, #4	Rhombus, p. 137

More about Assessment

Scoring and Analyzing Assessment Responses

Students may respond to assessment questions with various levels of mathematical sophistication and elaboration. Each student's response should be considered for the mathematics that it shows, and not judged on whether or not it includes an expected response. Responses to some of the assessment questions may be viewed as either correct or incorrect, but many answers will need flexible judgment by the teacher. Descriptive judgments related to specific goals and partial credit often provide more helpful feedback than percentage scores.

Openly communicate your expectations to all students, and report achievement and progress for each student relative to those expectations. When scoring students' responses, try to think about how they are progressing toward the goals of the unit and the strand.

Student Portfolios

Generally, a portfolio is a collection of student-selected pieces that is representative of a student's work. A portfolio may include evaluative comments by you or by the student. See the *Teacher Resource and Implementation Guide* for more ideas on portfolio focus and use.

A comprehensive discussion about the contents, management, and evaluation of portfolios can be found in *Mathematics Assessment: Myths, Models, Good Questions, and Practical Suggestions*, pp. 35–48.

Student Self-Evaluation

Self-evaluation encourages students to reflect on their progress in learning mathematical concepts, their developing abilities to use mathematics, and their dispositions toward mathematics. The following examples illustrate ways to incorporate student self-evaluations as one component of your assessment plan.

- Ask students to comment, in writing, on each piece they have chosen for their portfolios and on the progress they see in the pieces overall.

- Give a writing assignment entitled "What I Know Now about [a math concept] and What I Think about It." This will give you information about each student's disposition toward mathematics as well as his or her knowledge.

- Interview individuals or small groups to elicit what they have learned, what they think is important, and why.

Suggestions for self-inventories can be found in *Mathematics Assessment: Myths, Models, Good Questions, and Practical Suggestions*, pp. 55–58.

Summary Discussion

Discuss specific lessons and activities in the unit—what the student learned from them and what the activities have in common. This can be done in whole-class discussion, in small groups, or in personal interviews.

Connections across the *Mathematics in Context* Curriculum

Triangles and Beyond is the sixth unit in the geometry strand. The map below shows the complete *Mathematics in Context* curriculum for grade 7/8. This indicates where the unit fits in the geometry strand and in the overall picture.

A detailed description of the units, the strands, and the connections in the *Mathematics in Context* curriculum can be found in the *Teacher Resource and Implementation Guide.*

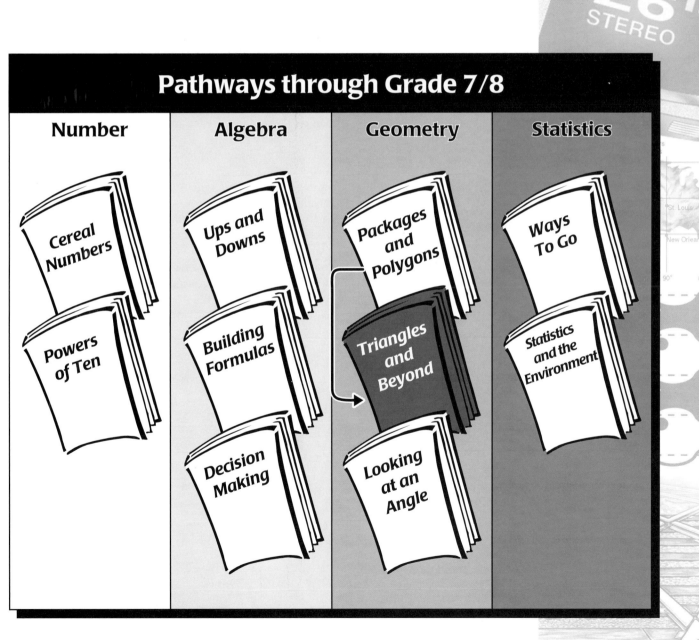

Pathways through Grade 7/8

Number	Algebra	Geometry	Statistics
Cereal Numbers	Ups and Downs	Packages and Polygons	Ways To Go
Powers of Ten	Building Formulas	Triangles and Beyond	Statistics and the Environment
	Decision Making	Looking at an Angle	

Grade 5/6

Side Seeing

*Figuring All
the Angles*

Grade 6/7

Reallotment

*Made to
Measure*

Grade 7/8

*Packages and
Polygons*

Ways to Go

*Triangles and
Beyond*

*Looking at
an Angle*

Grade 8/9

*Triangles and
Patchwork*

*Going the
Distance*

*Digging
Numbers*

Connections within the Geometry Strand

On the left is a map of the geometry strand; this unit, *Triangles and Beyond,* is highlighted.

Triangles and Beyond is the sixth unit in the geometry strand, and it is the second geometry unit at the grade 7/8 level, following the unit *Packages and Polygons.* Geometry in *Mathematics in Context* is presented through the concept of grasping space, to help students represent and make sense of the world. Understanding shapes is one of three substrands of the geometry strand. The exploration of space in the units in this strand and the more traditional geometry of forms and shapes blend together in *Packages and Polygons, Triangles and Beyond, Triangles and Patchwork* and *Going the Distance.* In the unit *Triangles and Beyond,* students explore triangles. Students learn to recognize triangles in the world around them, and learn to name different types of triangles. They investigate the possibility of constructing triangles with given side lengths and conclude this investigation by formulating a rule about the relationship among side lengths of a triangle. Another rule students develop in this unit involves the sum of the angles in a triangle. Students identify the relationship between the side lengths and the angles opposite them. Students also learn how to construct triangles and parallel lines and they identify properties of parallelograms. A final topic introduced in this unit is transformations: Students learn to identify and carry out translations, rotations, and reflections of figures. They rotate triangles to make polygons and parallelograms.

The Geometry Strand

Grade 5/6

Side Seeing
Exploring the relationship between three-dimensional shapes and drawings of them; seeing from different points of view, and building structures from drawings.

Figuring All the Angles
Estimating and measuring angles; and investigating direction, vectors, and rectangular and polar coordinates.

Grade 7/8

Packages and Polygons
Recognizing geometric shapes in real objects and representations; constructing models; and investigating properties of regular and semi-regular polyhedra.

Looking at an Angle
Recognizing vision lines in two and three dimensions; identifying and drawing shadows and blind spots; identifying the isomorphism of vision lines, light rays, flight paths, and so forth; understanding the relationship between angles and the tangent ratio; and computing with the tangent ratio.

Ways to Go
Reading and interpreting different kinds of maps; comparing different types of distances; progressing from one two-dimensional model to another (from a diagram to a map to a photograph to a graph); and drawing graphs and networks. (Ways to Go is also in the statistics strand.)

Triangles and Beyond
Exploring the interrelationships of the sides and angles of triangles as well as the properties of parallel lines and quadrilaterals; constructing triangles; and using transformations to become familiar with the concepts of congruence and similarity.

Grade 6/7

Reallotment
Measuring regular and irregular areas; discovering links between area, perimeter, surface area, and volume; and using English and metric units.

Made to Measure
Measuring length (including circumference), volume, and surface area using metric units.

Grade 8/9

Triangles and Patchwork
Understanding similarity and using it to find unknown measurements for similar triangles, and developing the concept of ratio through tessellation.

Going the Distance
Using the Pythagorean theorem to investigate distances, scales, and vectors; and using slope, tangent, area, square root, and contour lines.

Digging Numbers
Using the properties of height, diameter, and radius to determine whether or not various irregular shapes are similar; predicting length using graphs and formulas; exploring the relationship between three-dimensional shapes and drawings of them; and using length-to-width ratios to classify various objects. (Digging Numbers is also in the statistics strand.)

Connections with Other *Mathematics in Context* Units

Triangles and Beyond is a grade 7/8 unit in the geometry strand about understanding shapes. The theme of shape and construction is the major focus of this unit, which concentrates on triangles. This unit builds on ideas of shape learned in earlier units such as *Reallotment* and *Packages and Polygons*. It also builds on knowledge students acquired in the grade 5/6 unit *Figuring All the Angles* (where students learned to measure angles and turns). Ideas about measurement (for example, of angles) are also used in the grade 6/7 unit *Made to Measure*. However, this unit is not considered prerequisite to the unit *Triangles and Beyond*.

In *Triangles and Beyond*, the study of angle-side relationships in a triangle is formalized. The study of shape is continued in the grade 8/9 unit *Triangles and Patchwork*. There is also a connection to the grade 7/8 unit *Ways to Go*, in which students make constructions.

The following mathematical topics that are included in the unit *Triangles and Beyond* are introduced or further developed in other *Mathematics in Context* units.

Prerequisite Topics

Topic	Unit	Grade
angles	*Figuring All the Angles*	5/6
	Packages and Polygons	7/8
triangles and polygons	*Reallotment*	6/7
	Packages and Polygons	7/8
spatial visualization	*Side Seeing*	5/6

Topics Revisited in Other Units

Topic	Unit	Grade
angles	*Packages and Polygons*	7/8
	Looking at an Angle	7/8
	Made to Measure	6/7
triangles and polygons	*Packages and Polygons*	7/8
	Triangles and Patchwork	8/9
	Going the Distance	8/9
spatial visualization	*Looking at an Angle*	7/8
constructions	*Ways to Go**	7/8

* This unit in the statistics strand also helps students make connections to ideas about geometry.

Student Materials and Teaching Notes

Student Book
Table of Contents

Dear Student,

Welcome to *Triangles and Beyond.*

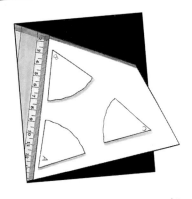

Throughout this unit, you will study many kinds of triangles and parallelograms and their special geometric properties.

What happens if you try to make a triangle with three straws, one 12 centimeters long, one 4 centimeters long, and one 6 centimeters long? Do you think there is any connection between the location of the largest angle and the location of the longest side in a triangle? As you work with triangles, you will learn some requirements for the measures of their sides and angles.

You will also investigate the properties of parallel lines and learn the differences between parallelograms, rectangles, rhombuses, and squares.

As you work through this unit, look around you to see how the geometric shapes and properties you are studying appear in everyday objects.

Sincerely,

The Mathematics in Context Development Team

Work Students Do

Students make a list of all the triangles they can see in the classroom. Then they locate other triangles in the world around them, looking for triangles in shapes, sculptures, houses, company logos, and three-dimensional objects. At the end of the section, students are shown triangles with parts missing, and they investigate how many ways such triangles can be completed.

Goals

Students will:

- recognize triangles in the world;
- appreciate that geometry is a means of describing the world;
- create mathematical representations from visualization.*

  * *This goal is introduced in this section and is assessed in other sections of the unit.*

Pacing

- approximately three 45-minute class sessions

About the Mathematics

This section offers students opportunities to locate triangles in illustrations and in the real world and to imagine what they would look like from different perspectives. This ability to visualize triangles will help students when they begin to study trigonometry.

Students look at triangles with some parts missing and determine which of these triangles can be reconstructed in only one way. Here, students are investigating the defining characteristic of any triangle—the relationship of the three sides to one another. If one side of a triangle is missing (see illustration below), there are infinite possibilities for its placement.

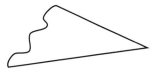

Materials

- Letter to the Family, page 124 of the Teacher Guide (one per student)
- Student Activity Sheets 1–3, pages 125–127 of the Teacher Guide (one of each per student)
- triangular objects or pictures of triangles, page 7 of the Teacher Guide, optional (several per class)
- newspapers and magazines, pages 7 and 9 of the Teacher Guide (several per class)
- toothpicks, page 9 of the Teacher Guide, optional (20 per student)
- clay or gumdrops, page 9 of the Teacher Guide, optional (10 per student)
- transparency of Student Activity Sheet 1, page 9 of the Teacher Guide, optional (one per class)
- other pictures of houses, page 11 of the Teacher Guide, optional (several per student)
- transparencies, pages 11 and 15 of the Teacher Guide, optional (one per class)
- colored overhead pens, pages 9, 11, 13, and 15 of the Teacher Guide, optional (two colors per class)
- overhead projector, pages 9, 11, 13, and 15 of the Teacher Guide, optional (one per class)
- drawing paper, page 11 of the Teacher Guide (one sheet per group of students)
- scissors, pages 11 and 15 of the Teacher Guide (one pair per student or group of students)
- tape, page 11 of the Teacher Guide (one roll per group of students)
- transparency of Student Book page 4, page 13 of the Teacher Guide, optional (one per class)
- examples of real company logos, page 13 of the Teacher Guide, optional (several per class)
- colored pens, page 15 of the Teacher Guide (three per student)
- transparent paper, page 15 of the Teacher Guide, optional (one sheet per student)
- straightedges or rulers, page 17 of the Teacher Guide (one per student)

Planning Instruction

Before beginning this section, you might want to bring several triangular objects into the classroom. Ask students to identify the triangles in the classroom and to think of some common objects that are shaped like triangles.

Students may work on problems 4, 5, 7, and 8 in small groups. They may do the remaining problems individually.

There are no optional problems in this section.

Homework

Problem 2 (page 8 of the Teacher Guide) can be assigned as homework. Also, the Extensions (pages 11 and 13 of the Teacher Guide) and the Bringing Math Home activity (page 7 of the Teacher Guide) can be assigned as homework. After students complete Section A, you may assign appropriate activities from the Try This! section, located on pages 45–48 of the *Triangles and Beyond* Student Book. The Try This! activities reinforce the key mathematical concepts introduced in this section.

Planning Assessment

- Problem 2 can be used to informally assess students' ability to recognize triangles in the world and to appreciate that geometry is a means of describing the world.

Finding Triangles

If you look around your classroom, you will probably see several triangles.

1. Make a list of all the triangles you see in the classroom.

Triangles can be found in many places. For instance, you might see triangles in the frame of a kite, a jungle gym, or a geodesic dome.

1. Answers will vary. Sample responses:

- the triangle formed by the hands of a clock

- a closed three-ring binder when looked at from the top

- triangles formed by a diagonal bar at the back of a bookcase (to stabilize an open bookcase)

- the projection mirror in the top part of an overhead projector when looked at from the side

- the tip of a pencil

- pictures of triangles on the cover of the Triangles and Beyond Student Book

Materials triangular objects or pictures of triangles, optional (several per class); newspapers and magazines, optional (several per class)

Overview Students identify triangles in the classroom.

Planning Students may work on problem **1** individually. When they are done, you may want to make a class inventory and discuss their answers.

Comments about the Problems

1. If the classroom does not contain many triangles, ask students *What other common objects have parts that are shaped like triangles?* [gables or windows on houses, spokes on wheels, pieces in kaleidoscopes, faces of pyramids] If students have difficulty thinking of examples, you might have them look for pictures of triangles in newspapers and magazines.

Some students may identify objects that have an implied triangular shape, such as the two hands of a clock, as shown below.

If students suggest a clock, you might ask, *Do the hands always make a triangle?* [No, the hands do not make a triangle when they make angles of 0° and 180°.]

Bringing Math Home You might have students look for triangles at home or in other places. They may involve their families in their search for triangles.

Did You Know? Buckminster Fuller, an American engineer and architect, developed the geodesic dome. The geodesic dome is a self-supporting structure. No columns are needed to support the roof no matter how large it is. Fuller proposed that geodesic domes be used as "sky breaks" over entire cities. Inside the dome, people could control their environment, even in the Arctic or Antarctic.

Activity

2. Find some other examples of triangular objects in photographs or pictures from magazines and newspapers. Paste your pictures in your notebook or make a poster or a collage with them. Save your examples to use later in this unit.

Shown below is a photograph of a bridge over the Rio Grande near Santa Fe, New Mexico. Next to the photograph is a drawing of one section of the bridge.

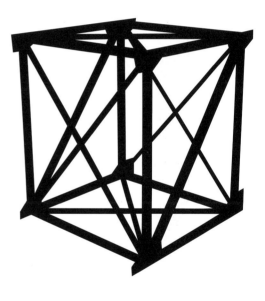

Some of the triangles you see in the drawing are actually part of the bridge. Other triangles can only be seen in the drawing.

3. On **Student Activity Sheet 1,** use one color to show three triangles that are part of the bridge's frame. Use another color to show three triangles that can only be seen in the drawing. You can check your answer by making a model of the bridge using toothpicks and clay or gumdrops.

2. Any drawings and pictures of triangular objects or of things containing triangular shapes are acceptable.

3. Answers will vary. Sample response:

Triangles That Are Part of the Bridge

Triangles Only Visible in the Drawing

For the triangles that are part of the bridge, there are eight triangles on top, four small and four large, and eight on the bottom as shown below.

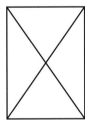

There are two triangles in the front, and two in the back as shown below.

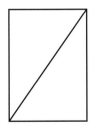

There are eight triangles at each of the two other sides, as shown below.

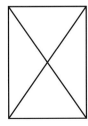

All together, there are 16 + 4 + 16 = 36 triangles.

Materials Student Activity Sheet 1 (one per student); toothpicks, optional (20 per student); clay or gumdrops, optional (10 pieces per student); transparency of Student Activity Sheet 1, optional (one per class); colored overhead pens, optional (two colors per class); overhead projector, optional (one per class); newspapers and magazines (several per class)

Overview Students find pictures of triangles in magazines or newspapers. They then paste the pictures in their notebooks or make a poster with them.

Planning Students may work on problems **2** and **3** individually. Problem **2** may be used as informal assessment. Discuss students' answers in class.

Comments about the Problems

2. **Informal Assessment** This problem assesses students' ability to recognize triangles in the world. You may want to have students make a collage or use notebooks to keep track of their collections. Have students save their work, because they will need it later. This problem can also be assigned as homework.

3. If students have difficulty visualizing the triangles in the bridge's frame, you might have them construct a model using toothpicks and clay or gumdrops. You may also want to prepare a copy of Student Activity Sheet 1 on a transparency to use in your class discussion of this problem. Encourage students to label vertices with letters. They can then identify each triangle with the three letters corresponding to each of the triangle's vertices.

Explain that although the actual bridge is three-dimensional, the drawing is two-dimensional. As a result, some bars that do not actually touch each other appear to touch each other and form triangles in the drawing. But a real triangle that is part of the actual bridge's construction is formed by three sides that meet at three vertices.

Make a flag similar to the one on the right (a triangular piece of paper taped to a pencil will work). Hold the flag at eye level and slowly rotate it.

4. In your notebook, make several drawings showing the changes that you see as you rotate your flag.

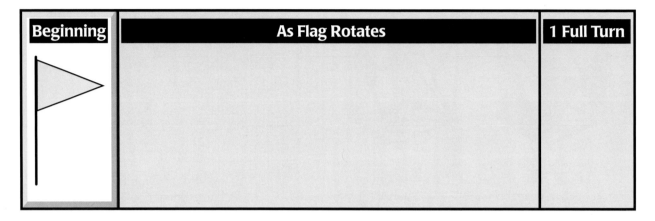

Beginning	As Flag Rotates	1 Full Turn

Some houses have slanted roofs, such as the one shown below. Slanted roofs produce some interesting triangles.

5. a. Count the number of triangles that you can see in the drawing of the house.

b. Do you think there are any triangles on the house that you cannot see in the drawing? Explain.

Sometimes you cannot see the actual shapes of the triangles and other objects in a drawing because of the perspective of the drawing.

6. a. Sketch the front view of the house. Pay attention to the shape of the triangular gable (the pitched roof above the front door).

b. Why is the shape of the front triangle on the gable different in your drawing than it is in the above drawing?

4. Drawings will vary. Sample drawings:

Beginning　　　　**As Flag Rotates**　　　　**1 Full Turn**

5. a. Answers will vary. Sample response:

There are six triangles, as shown in the drawing below:

b. Yes. There must be at least one other triangle for the other peak of the roof.

6. a. Drawings will vary. Sample drawing:

b. Answers will vary. Sample student response:

The side lengths and the angle measurements in the two triangles are different because the sketch in the book is drawn at an angle, while my drawing is from directly in front of the house.

Materials drawing paper (one sheet per group of students); scissors (one pair per group of students); tape (one roll per group of students); transparency, optional (one per class); overhead projector, optional (one per class); colored overhead pens, optional (two colors per class); other pictures of houses, optional (one per student)

Overview Students construct a flag and draw different views of it as it is rotated. They identify triangles that can be seen in the drawing of a house, and reason about triangles that cannot be seen. They also investigate distortion due to perspective.

About the Mathematics Students learned to recognize, draw, and reason about different views of objects in the grade 5/6 unit *Side Seeing*.

Planning Students may work on problems **4** and **5** in small groups. Problem **6** can be done individually.

Comments about the Problems

4. Students may share the work in small groups. For example, one student may hold the flag and rotate it while the others draw the corresponding views. If students have difficulty, you might discuss the difference between a full turn, a half turn, and a quarter turn. For example, after a full turn, the flag should be in its original position. After a half turn, the flag should point in the opposite direction.

5. a. Students should discuss the different triangles they see.

b. Students may want to visualize what the house would look like from the other side.

6. a. *Gable* may be a new word for some students. A gable is the triangular end of a sloped roof.

Extension You may wish to have students bring in other pictures of houses. Then, discuss how drawings or photographs may distort the actual dimensions of a house. You might have students experiment with visualizing distorted figures by drawing shapes on a transparency, projecting them on the overhead, and having students look at the screen from different places in the room. You can also hold the transparency at an angle to the overhead.

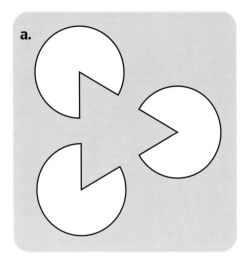

a.

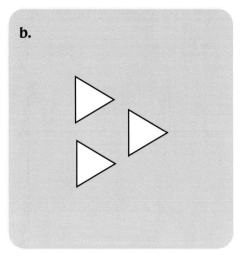

b.

7. How many triangles do you see in each of these drawings? Explain.

Triangles show up in many different places. Some company logos that use triangles are shown below.

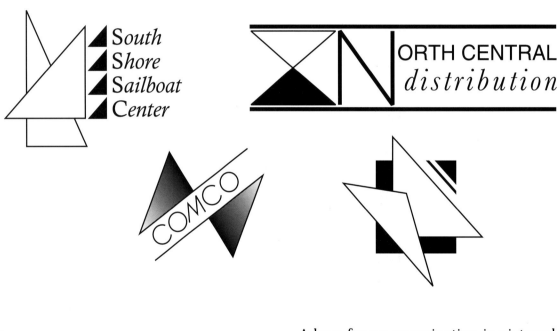

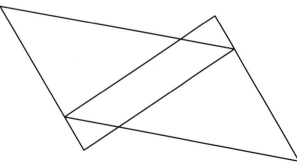

A logo for an organization is pictured on the left.

8. a. How many triangles do you see in this logo?

b. What do you think this logo symbolizes?

7. Answers will vary. Sample responses:

a.

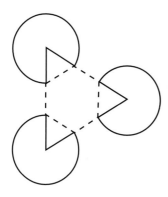

- zero, since a real triangle is not shown
- one, if you count the illusion of a triangle formed by the circles
- three, if you just count implied triangles formed by wedges in the circles
- four, if you count the illusion of one large triangle plus the three smaller triangles formed by the wedges in the circles

b. Answers will vary. Sample responses:

- three, if only real triangles are included

- four, if the real triangles are considered to form the corners of a larger triangle or if they are interpreted as bordering a triangle in the middle

- five, if the three real triangles, the implied triangle in the middle, and a large triangle formed by all four small triangles are included

8. a. There are six triangles total, as shown below: two large interlocking triangles, two small triangles at opposite corners, and two medium triangles in the other corners.

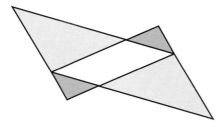

b. Answers will vary. Students may respond that the interlocking triangles show togetherness or cooperation.

Materials colored overhead pens, optional (two colors per class); overhead projector, optional (one per class); transparency of of Student Book page 4, optional (one per class); examples of real company logos, optional (several per class)

Overview Students look for triangles in company logos.

Planning Students may work on problems **7** and **8** in small groups. Discuss their answers as a class.

Comments about the Problems

7. You may wish to copy this Student Book page onto a transparency and draw the triangles with colored pens as students discuss them. This problem may again lead to some discussion about the difference between real and implied triangles. You may wish to ask students what defines a triangle and keep a list of their answers for discussion later in the unit.

8. Explain to students that a logo is a symbol that represents a company or an organization. You might bring in the school letterhead or newsletter to demonstrate.

If students have difficulty identifying the triangles in this logo, you might have them label all the points of intersection and name each triangle individually.

Extension You may wish to have students make a list or collage of real logos that incorporate triangles. (The logos on this page have been created for this unit.)

The figure on the left is an outline of three different triangles that have been stacked on top of each other.

9. a. On **Student Activity Sheet 2,** use different colors to outline each of the three triangles in the top picture.

   b. In how many ways can the three triangles be stacked? Color the figures on **Student Activity Sheet 2** to show all the possibilities. (You may not need all the figures.)

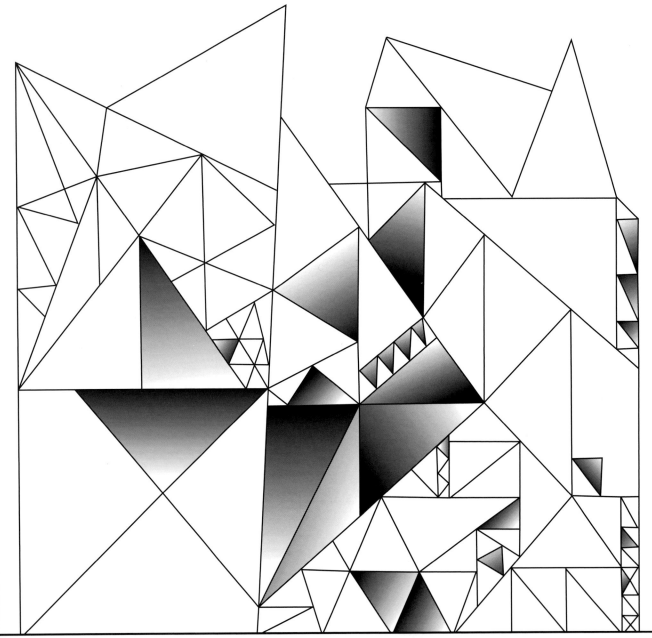

Solutions and Samples
of student work

9. a.

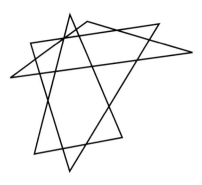

b. There are six possibilities. They are outlined in the pictures below.

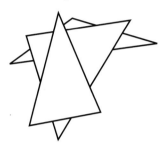 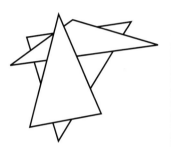

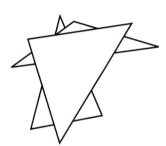 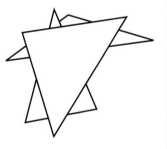

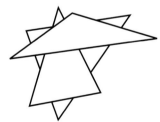 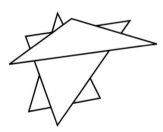

Materials Student Activity Sheet 2 (one per student); colored overhead pens, optional (two colors per class); overhead projector, optional (one per class); transparency, optional (one per class); colored pens (three per student); transparent paper, optional (one sheet per student); scissors, optional (one pair per student)

Overview Students identify triangles in an outline of three triangles. They also investigate the different ways in which three triangles may be stacked.

About the Mathematics Problem **9b** is a permutation problem: it deals with finding all possible orderings, given a certain number of objects. Another example of such an ordering problem would be finding all possible ways in which four different students can sit in four desks.

It works best to find the possibilities in a systematic way. For example, four different students can sit in the first desk, but after one person sits in the first desk, there are only three possibilities for the second desk, two possibilities for the third desk, and only the fourth person can occupy the fourth desk. This yields 4 × 3 × 2 × 1 = 24 different ways.

Planning Students may work on problem **9** individually. You may want to discuss students' answers using a transparency and colored overhead pens. You might also make a display of students' work.

Comments about the Problems

9. It is more important that students are able to visualize the triangles than it is for them to properly order them.

b. If students have difficulty, you might have them list all the possibilities before they start coloring. You can also do one example as a class.

Encourage students to work systematically. For example, they might place one triangle on top and find all possibilities. Then they can place another triangle on top.

Students should color whole triangles, not just the outlines. That way, they can clearly see which triangle is on top, which is in the middle, and which is on the bottom. If they still have difficulty visualizing the triangles, you might have them trace each triangle individually, cut the triangles out, color them, and stack them.

Summary

Triangles appear in many places. They may be part of a building's structure, a company's logo, or the shape of a flag.

Depending on the perspective from which you view a triangle, its shape can vary.

Summary Questions

The figures below are triangles that have been torn.

A.

B.

C.

D.

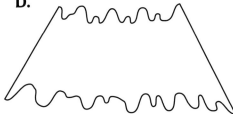

10. a. On **Student Activity Sheet 3,** draw the missing parts to restore the triangles to their original shapes.

    b. Which of the triangles can be restored in more than one way? Why do you think only some of the triangles can be restored in more than one way?

10. a. Drawings will vary, depending on how far students extend the sides of triangles C and D. Sample drawings:

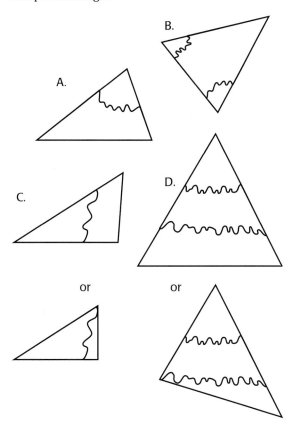

b. Triangles C and D can be restored in numerous ways. When the sides are approaching each other (as in triangles A and B), there is only one possibility. When the lines are spreading out (as in triangles C and D), there are many ways to join the sides. Another way of looking at this is to see that the bottom two triangles have parts of only two sides given. The top two triangles have parts of all three sides given.

Materials Student Activity Sheet 3 (one per student); straightedges or rulers (one per student)

Overview Students read the Summary, which reviews the main concepts covered in this section. Then they restore torn triangles and describe why some may be completed in more than one way.

Planning Students may work on problem **10** individually. Discuss their answers in class. After students complete Section A, you may assign appropriate activities from the Try This! section, located on pages 45–48 of the *Triangles and Beyond* Student Book, as homework.

Comments about the Problems

10. a. Investigate whether students have completed the triangles in different ways, especially for triangles C and D. Have students show different solutions to the class, and let the class decide whether they are correct.

   b. Some students may find their solutions by drawing the possibilities, while others may reason about how the triangles can be restored just by looking at the segments.

   Ask students, *What is the most damage you could do to a triangle and still recover the original triangle?* [You can recover the original triangle if you are given parts of all three sides, as shown below, because a triangle is formed by three nonparallel lines. If you do not have a part of each side, you cannot recover the triangle.]

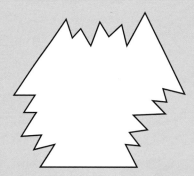

Work Students Do

Students form triangles out of uncooked spaghetti noodles broken into specified lengths. In this way, they are introduced to scalene, isosceles, and equilateral triangles. Students then categorize triangles they find in photos and illustrations as being one of the three types. In the context of a game of Frisbee between three people, students investigate how to construct a triangle with given side lengths. Given a line that represents the distance of the second player from the first, students then draw arcs from the endpoints that represent the distances of the third player from the first and from the second. Where these arcs intersect is the location of the third player and the third point of the triangle. Students discover why it is important to use a compass in constructing a triangle. Finally, students use formal construction techniques to make equilateral and isosceles triangles.

Goals

Students will:

- use properties of triangles (such as the sum of angles, side relationships, and the Hinge theorem);
- construct a triangle with given side lengths;
- use construction when appropriate;
- describe geometric figures using words and/or diagrams.

Pacing

- approximately three 45-minute class sessions

Vocabulary

- equilateral triangle
- isosceles triangle
- scalene triangle

About the Mathematics

This section introduces the characteristic of triangles that no one side can be longer than the sum of the lengths of the other two sides. Triangles are defined as being one of three types: equilateral, isosceles, or scalene. Equilateral triangles have all three sides and all three angles equal. Isosceles triangles have at least two sides and two angles equal. And scalene triangles have no equal sides or angles.

Materials

- Student Activity Sheets 4 and 5, pages 128–129 of the Teacher Guide (one of each per student)
- uncooked, dry spaghetti, page 21 of the Teacher Guide (four strands per student)
- crayons or markers, page 21 of the Teacher Guide, optional (three colors per student)
- sets of spaghetti pieces, made by students on page 7 of the Student Book, pages 23 and 25 of the Teacher Guide (four sets per student)
- pictures of triangles gathered by students on page 2 of the Student Book, page 25 of the Teacher Guide (several per student)
- centimeter rulers, pages 27, 31, 33, and 35 of the Teacher Guide (one per student)
- string, page 29 of the Teacher Guide (about 1–2 meters per group of students)
- compasses, pages 29, 31, and 35 of the Teacher Guide (one per student)
- transparency of an isosceles triangle, page 33 of the Teacher Guide, optional (one per class)
- overhead projector, page 33 of the Teacher Guide, optional (one per class)
- large piece of cardboard, page 33 of the Teacher Guide, optional (one per class)

Planning Instruction

You may want to begin this unit by discussing ways in which triangles can be useful in real life. For example, triangles are often used in construction because they are strong and stable. Point out to students that if they sit with their legs crossed (one foot under each thigh), they will be stable because their legs will form a triangle. Properties of triangles can also be used to solve problems without measuring. Ask students, *What triangles in real life have sides or angles that would be difficult to measure?* [the flight path of a plane taking off or landing, triangles formed by the stars in the sky and by landmarks on Earth, etc.

Students may work on problems 1a, 8b, 12, 14, and 15 individually. They may do problems 3, 4, and 11 individually or in small groups. The remaining problems should be done in small groups.

There are no optional problems in this section.

Homework

Problem 5 (page 24 of the Teacher Guide) may be assigned as homework. After students complete Section B, you may assign appropriate activities from the Try This! section, located on pages 45–48 of the *Triangles and Beyond* Student Book. The Try This! activities reinforce the key mathematical concepts introduced in this section.

Planning Assessment

- Problem 5 can be used to informally assess students' ability to describe geometric figures using words and/or diagrams.
- Problem 12 can be used to informally assess students' ability to use properties of triangles (such as the sum of angles, side relationships, and the Hinge Theorem) and construct a triangle with given side lengths.
- Problem 14 can be used to informally assess students' ability to describe geometric figures using words and/or diagrams, and to construct a triangle with given side lengths.
- Problem 15 can be used to informally assess students' ability to use properties of triangles (such as the sum of angles, side relationships, and the Hinge Theorem) and to use construction when appropriate.

Activity

MAKING TRIANGLES

For this activity, you will need four long pieces of uncooked, dry spaghetti. Carefully break one of your spaghetti pieces into the lengths shown for set A. Break the others to match the lengths for sets B, C, and D.

Set A
3 cm
4 cm
5 cm

Set B
3 cm
3 cm
7 cm

Set C
3 cm
5 cm
7 cm

Set D
3 cm
4 cm
7 cm

Students' spaghetti pieces should match the ones shown on page 7 of the Student Book.

Materials uncooked, dry spaghetti (four strands per student); crayons or markers, optional (three colors per student)

Overview Students break pieces of spaghetti to match lengths that are given. They will use these sets of spaghetti strands to make triangles. There are no problems for students to solve on this page.

Planning Students may begin the activity individually. Each student should make each set of spaghetti lengths. You may wish to have students color the tips of the pieces of spaghetti in each set a different color. Save the sets of spaghetti strands for use later in this section.

	Sides (in cm)	Sketch of What Happened	Can You Make a Triangle? Explain.
Set A	3, 4, 5		
Set B	3, 3, 7		
Set C	3, 5, 7		
Set D	3, 4, 7		

1. a. Try to make a triangle with the three lengths in each set. Then copy and complete the above chart.

   b. In order to form a triangle, there is a requirement for the lengths of the three sides. Describe this requirement in your own words.

   c. Describe the angles of each triangle that you were able to make.

Triangles are divided into three different categories according to the lengths of their sides:

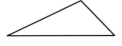

- Triangles with three sides of different lengths are called *scalene* triangles.

- Triangles with at least two equal sides are called *isosceles* triangles.

- Triangles with three equal sides are called *equilateral* triangles.

1. a. Sketches and explanations will vary. Sample table:

Sides (in cm)	Sketch of What Happened	Can You Make a Triangle? Explain.
Set A 3, 4, 5		yes
Set B 3, 3, 7		no, the sides didn't reach
Set C 3, 5, 7		yes
Set D 3, 4, 7		no, the triangle was flat

b. Answers will vary. Sample response:

The sum of any two sides of a triangle must be greater than the length of the third side.

c. Answers will vary. Sample responses:

Set A made a triangle with a right angle and two smaller angles.

Set C had one angle larger than a right angle and two angles smaller than a right angle.

Materials sets of spaghetti strands, made by students on page 7 of the Student Book (four sets per student)

Overview Students try to create triangles using the sets of spaghetti pieces they made. They discover a requirement for the lengths of the three sides of a triangle. They also learn the names of different types of triangles.

About the Mathematics Given three line segments, the length of any one side must be shorter than the sum of the lengths of the other two sides in order to form a triangle.

Planning Students may complete problem **1a** individually before discussing problems **1b** and **1c** in small groups.

Comments about the Problems

1. a. Encourage students to sketch the results of their trials.

b. Students should discover that a triangle cannot be made out of any three line segments. Instead, the length of any one side must be shorter than the sum of the lengths of the other two sides. Otherwise, the triangle either is flat or cannot be closed at all. You may want to discuss students' answers as a class and formulate a class definition for this rule.

c. Some students may remember obtuse and acute triangles from *Figuring All the Angles.*

Extension Ask students, *Is it possible to draw a triangle that is not scalene, isosceles, or equilateral?* [no]

Activity

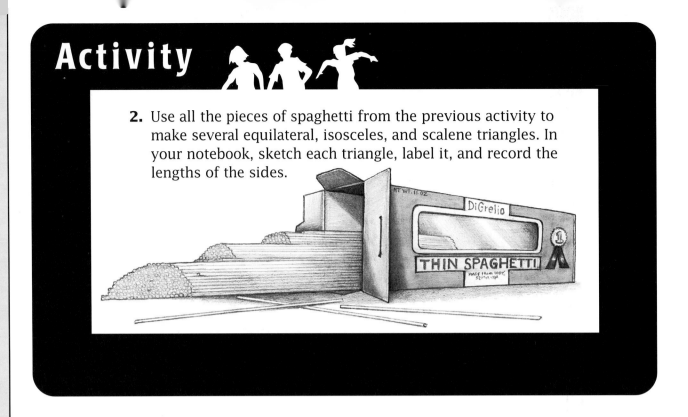

2. Use all the pieces of spaghetti from the previous activity to make several equilateral, isosceles, and scalene triangles. In your notebook, sketch each triangle, label it, and record the lengths of the sides.

Looking at the Sides

3. What kind of triangles form the geodesic dome on page 1?

4. Make an isosceles triangle without measuring by folding a strip of paper or a drinking straw. How can you be sure that your triangle is isosceles?

5. At the beginning of Section A, you gathered some pictures of triangles. Which of your examples are isosceles triangles? Which are equilateral? Which are scalene?

2. Possible equilateral triangles:
- 3 cm, 3 cm, 3 cm
- 7 cm, 7 cm, 7 cm

Possible isosceles triangles:
- 3, 3, 3 • 4, 4, 3 • 5, 5, 3 • 7, 7, 3
- 3, 3, 4 • 4, 4, 5 • 5, 5, 4 • 7, 7, 4
- 3, 3, 5 • 4, 4, 7 • 5, 5, 7 • 7, 7, 5
                                  • 7, 7, 7

Possible scalene triangles:
- 3 cm, 4 cm, 5 cm
- 3 cm, 5 cm, 7 cm
- 4 cm, 5 cm, 7 cm

3. Answers may vary. They appear to be equilateral. Students might also say that they are equilateral along the bottom of the dome and isosceles further up.

4. Any isosceles triangle is acceptable, but students should be able to show how they made two sides equal.

Students may make an isosceles triangle in two steps. First, they should fold one part of the straw or strip in order to get two pieces of the same length, as shown below.

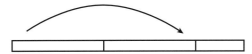

Second, students should bend the unequal piece to complete the triangle, as shown below.

folding
line

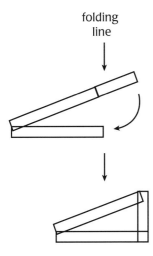

5. Answers will vary, but students should be able to justify their answers by showing that isosceles triangles have at least two equal sides, equilateral triangles have three equal sides, and scalene triangles have sides of three different lengths.

Materials sets of spaghetti strands, made by students on page 7 of the Student Book (four sets per student); pictures of triangles gathered by students on page 2 of the Student Book (several per student)

Overview Students make different types of triangles from spaghetti strands and label the triangles as equilateral, isosceles, or scalene.

Planning Students may work on problem **2** in small groups. Problems **3** and **4** can be done individually or in small groups before discussing them in class. Problem **5** may be done individually and used for informal assessment. Problem **5** may also be assigned as homework.

Comments about the Problems

4–5. These problems may lead to a discussion about the definitions of isosceles and equilateral triangles. Ask students, *Are all equilateral triangles isosceles? Are all isosceles triangles equilateral?* [Equilateral triangles are also isosceles, but not all isosceles triangles are equilateral.] One way to explain this is to use a tree diagram, as shown below.

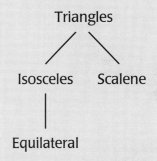

4. You might ask students, *How many isosceles triangles are possible?* [infinitely many]

5. Informal Assessment This problem assesses students' ability to describe geometric figures using words and/or diagrams. This problem can also be assigned as homework.

The Park

Anita *(A)*, Beth *(B)*, and Chen *(C)* want to play catch with a Frisbee in the park. To compensate for their different ages, they agree to stand at the positions shown in the diagram on the right (the arrows show the directions of the Frisbee throws).

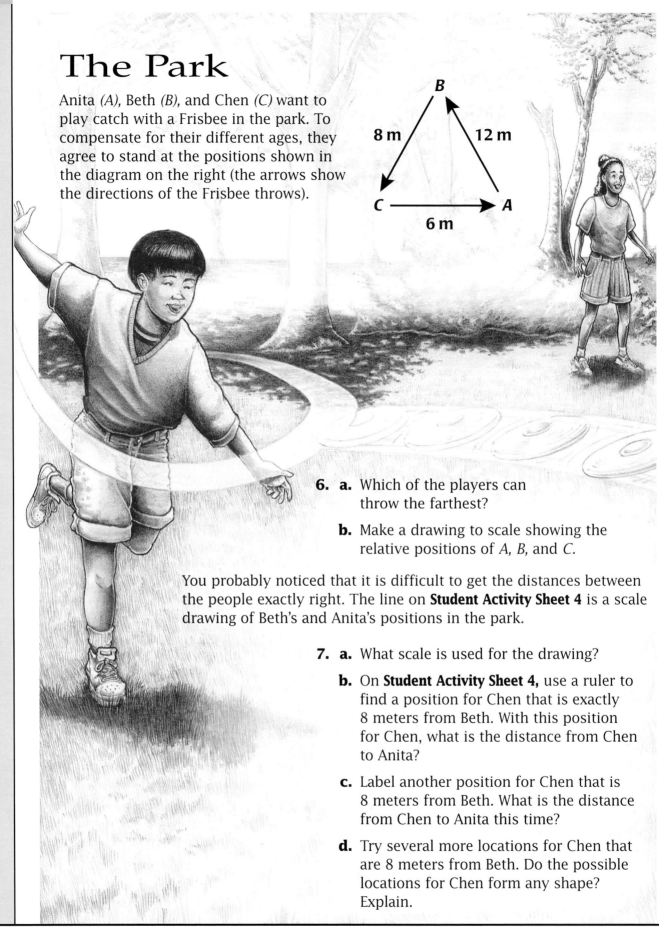

6. a. Which of the players can throw the farthest?

b. Make a drawing to scale showing the relative positions of *A*, *B*, and *C*.

You probably noticed that it is difficult to get the distances between the people exactly right. The line on **Student Activity Sheet 4** is a scale drawing of Beth's and Anita's positions in the park.

7. a. What scale is used for the drawing?

b. On **Student Activity Sheet 4,** use a ruler to find a position for Chen that is exactly 8 meters from Beth. With this position for Chen, what is the distance from Chen to Anita?

c. Label another position for Chen that is 8 meters from Beth. What is the distance from Chen to Anita this time?

d. Try several more locations for Chen that are 8 meters from Beth. Do the possible locations for Chen form any shape? Explain.

6. a. Player *A* (Anita)

b. Drawings will vary. Sample drawing:

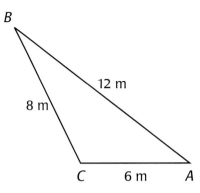

scale: 1 cm represents 2 m.

7. a. Each centimeter on the drawing represents one meter in the park.

b. Answers will vary. The distance from Chen to Anita could range from 4 to 20 meters.

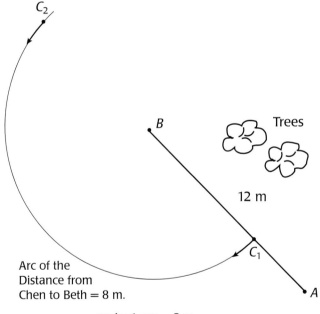

Arc of the
Distance from
Chen to Beth = 8 m.

scale: 1 cm = 2 m

c. Answers will vary. As in problem **7b,** the distance from Chen to Anita could range from 4 to 20 meters.

d. Yes. The points form a semicircle with a radius of 8 meters.

Materials Student Activity Sheet 4 (one per student); centimeter rulers (one per student)

Overview Students try to make a correct scale drawing of the relative positions of three Frisbee throwers with given distances between them.

About the Mathematics Students are slowly led to the idea of using a compass to construct a triangle given the lengths of the three sides.

Planning Students may work on problems **6** and **7** in small groups. Discuss their answers.

Comments about the Problems

6. b. The purpose of this problem is for students to learn how difficult it is to construct a triangle with exact measurements. Allow students to work long enough to understand the need for a better method than trial-and-error, but not so long that they become frustrated. Some students may remember this type of construction from the grade 7/8 unit *Ways to Go.*

7. a. Students should use a centimeter ruler to answer this question.

b. Some students may label points in the tree area, which is fine. Students should get a picture of all the possible positions for Chen.

d. This question is leading up to the compass construction of a triangle when the lengths of all three sides are known. Some students may note that a circle is the set of points that are a certain distance (radius) from a point.

Activity

With a partner, you can demonstrate how to find all possible locations for one point of the triangle. With each person holding an end of a long piece of string, stand as far apart from your partner as you can. Have one person stand still while the other walks in a path that keeps the string tight. (Keeping the string tight means that your distance apart remains the same.)

8. a. What is the shape of the path made by the person who is walking?

 b. Use a compass to make a drawing to scale of this path.

9. a. Use a compass to find all possible positions that are 8 meters from Beth on **Student Activity Sheet 4.**

 b. Now use a compass to draw all possible positions that are 6 meters from Anita on **Student Activity Sheet 4.**

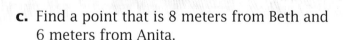

 c. Find a point that is 8 meters from Beth and 6 meters from Anita.

 d. If Anita, Beth, and Chen play in an area that is not blocked by trees on one side, how many possible locations are there for Chen?

Beth has to go home. Anita and Chen look for a new player, but this means they need a new triangle. Anita and Chen each want to stand the same distance apart as before.

10. a. Could a person who usually throws the Frisbee a distance of 20 meters play with them? Explain.

 b. What is the range of distances that the third player could throw to be in the game?

8. a. The shape of the path is a circle.

  b. Drawings will vary.

9. a–c.

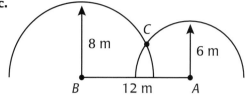

  d. There are two possible locations for Chen, as shown below.

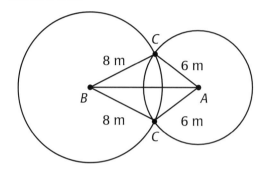

10. a. Answers may vary. Sample student responses:

> Yes, as long as the person is willing to throw less than the full 20 meters. Actually, the person would have to throw less than 18 meters to play.

> No, that is too far for any triangle with A and C. If the distance between C and A is 6 meters, and the distance between A and the new player is 12 meters, then the distance between the new player and C must be less than 6 + 12, or 18 meters. If the new player is 20 meters from Chen, Anita will not be able to throw far enough to reach him or her.

  b. anywhere between 6 and 18 meters, as shown below:

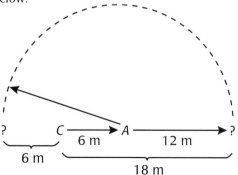

Materials Student Activity Sheet 4 (one per student); string (about 1–2 meters per group of students); compasses (one per student)

Overview Using string, students determine all possible locations that can be reached with a given throwing distance. They then use a compass to construct triangles in order to determine possible relative locations for Frisbee throwers who can throw given distances.

Planning The activity can be done in small groups or with the whole class. Students may work on problem **8b** individually. Problems **9** and **10** can be done in small groups. However, be sure to discuss problem **9** in class before students do problem **10**.

Comments about the Problems

8. You may want to have students practice drawing circles with a compass before doing this problem. Students should do this problem by walking, rather than just using string on paper, so that they get a strong spatial feel for a circle. Once students have walked, they can use string or a compass on the paper. *Note:* The compass is truly the equivalent of the string. A compass might just as well look like the following:

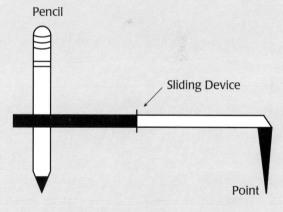

10. Students may solve this problem by sketching the possible locations for each player or by reasoning based on the rule for the lengths of sides of triangles.
If students have difficulty, you might encourage them to sketch the situation using a compass.

  b. Note that C cannot be exactly 6 meters or 18 meters away from A, because that would make a flat triangle. Points A, B, and C would be standing exactly in line.

Suppose the sides of a triangle have the following lengths: AB = 7 centimeters, AC = 5 centimeters, and BC = 6 centimeters.

11. a. Draw side AB on a blank piece of paper.

   b. From B, use a compass to find all possible locations of C.

   c. From A, use a compass to find all possible locations of C.

   d. In problem **1b** of the spaghetti activity, you found a requirement for making triangles. How does using a compass to make a triangle illustrate this requirement?

The following table lists sets of side lengths that may or may not make triangles.

	Length (in centimeters)		
	Side AB	Side BC	Side AC
Set 1	16	14	8
Set 2	24	10	12
Set 3	20	15	16
Set 4	13	7	21

12. Without drawing them, tell which sets in the above table form triangles. Construct one of the triangles on paper.

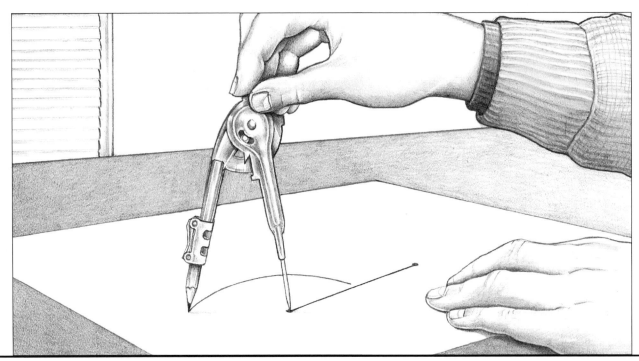

11. a–c.

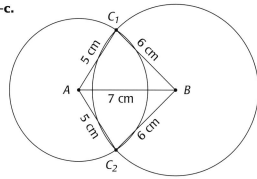

d. Explanations will vary. Sample explanation:

Two circles drawn at the endpoints need to be big enough so that they touch. They must touch in two places; if they touch in only one place, the sides are not long enough to form a triangle.

12. Sets **1** and **3** form triangles. Sets **2** and **4** do not form triangles because each has a side that is larger than the sum of the other two sides. Sample drawings:

Set 1

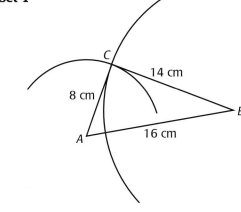

Set 3

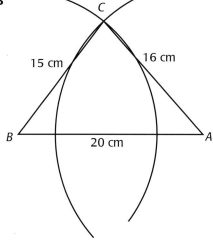

Materials centimeter rulers (one per student); compasses (one per student)

Overview Students construct a triangle with given side lengths. They determine, without drawing, which sets of given side lengths will form triangles.

Planning Students may work on problem **11** individually or in small groups. Discuss students' answers. Problem **12** may be done individually and used as an informal assessment.

Comments about the Problems

11. a–c. Students should focus on the triangle, not the circles. After some practice, they should need to make only short arcs instead of entire circles.

   d. If students have difficulty, you might remind them of the requirement they discovered in the spaghetti activity. Using a compass to construct triangles enables students to reflect on their earlier discovery.

12. Informal Assessment This problem assesses students' ability to use properties of triangles and construct a triangle with given side lengths.

If you are riding in a car and pass a triangular sign, such as the one in the photos shown below, you may notice that the shape of the triangle appears to change. (*Note:* The photos shown below are not in any particular order.)

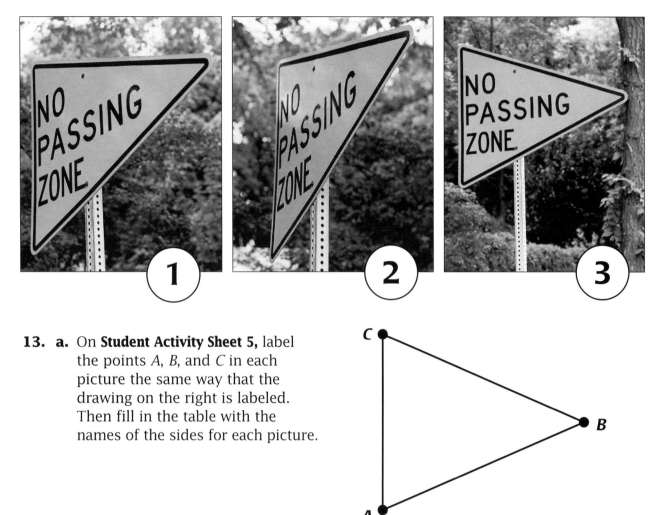

13. **a.** On **Student Activity Sheet 5,** label the points *A*, *B*, and *C* in each picture the same way that the drawing on the right is labeled. Then fill in the table with the names of the sides for each picture.

Photo	Longest Side	Middle Side	Shortest Side
1			
2			
3			

b. What kind of triangle do you see when you are standing directly in front of the sign?

c. If these photos were taken by someone driving toward the sign, in what order were they taken?

Solutions and Samples
of student work

13. a.

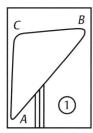 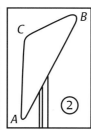 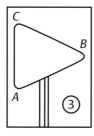

Photo	Longest Side	Middle Side	Shortest Side
1	AB	AC	BC
2	AB	AC	BC
3	AB	BC	AC

These lengths are very close.

b. isosceles

c. 3, 1, 2

Materials Student Activity Sheet 5 (one per student); centimeter rulers (one per student); transparency of an isosceles triangle, optional (one per class); overhead projector, optional (one per class); large piece of cardboard, optional (one per class)

Overview Students study the shape of a triangular sign and note how it changes when photographs of it are taken from different angles. Students put photographs in the order in which they were taken. Also, students label the longest, middle, and shortest sides.

About the Mathematics Being able to determine the shortest, middle, and longest sides is important to understanding the so-called Hinge Theorem, which states that in any triangle, the longest side is opposite the largest angle and the shortest side is opposite the smallest angle. This theorem is introduced in Section D.

Planning Students may work on problem **13** in small groups. Discuss students' answers in class. You may decide to do problem **13c** as a whole class.

Comments about the Problems

13. You may want to remind students of the activity with the flag on a pencil that they did on page 3 of the Student Book, since it is similar to problem **13c.** If students have difficulty, you might try placing an isosceles triangle on an overhead projector to demonstrate how it looks different to students located at opposite places in the room. You might also make a sign out of a large piece of cardboard, and hang it in the classroom. Students could walk past the sign, observing how it looks from different angles.

Summary

For any triangle, the sum of the lengths of any two sides is greater than the length of the remaining side.

$AB + BC > AC$

$AB + AC > BC$

$AC + BC > AB$

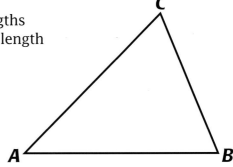

Using a compass and a straightedge, you can construct any triangle given three side lengths that meet the above condition.

A scalene triangle has three sides of different lengths.

An isosceles triangle has at least two sides of equal length.

An equilateral triangle has three sides of equal length.

Summary Questions

14. Construct an isosceles triangle and an equilateral triangle using only a compass and a straightedge, not a ruler.

15. Aaron wants to make a triangular structure using three beams. The longest beam is 10 centimeters. The other two beams are 4 centimeters each. Explain why Aaron cannot form a triangle with the three beams. Also explain how he can change the length of one of the beams to make a triangle.

14. Triangles will vary. Sample drawings:

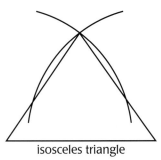

isosceles triangle

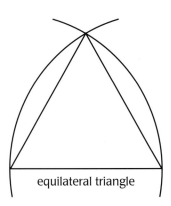

equilateral triangle

After drawing a line segment for an isosceles triangle, the compass setting must be the same at both endpoints. For an equilateral triangle, the compass setting at each endpoint must be the same length as the segment drawn for the first side.

15. Aaron cannot form a triangle with the three beams because the sum of the lengths of the two shortest beams is less than the length of the third beam, so they cannot meet to form a triangle.

Answers will vary as to how to make a triangle with the given three beams, but Aaron will have to cut the longest beam so that it is less than 8 centimeters long.

Materials centimeter rulers (one per student); compasses (one per student)

Overview Students read the Summary, which reviews the main concepts covered in this section. Then, students construct an isosceles triangle and an equilateral triangle. They explain why a third triangle is impossible to construct and describe how to change the length of one side to make it possible.

Planning Students may work on problems **14** and **15** individually. These problems may also be used for informal assessment. Discuss students' answers in class. After students complete Section B, you may assign appropriate activities from the Try This! section, located on pages 45–48 of the *Triangles and Beyond* Student Book, for homework.

Comments about the Problems

14. Informal Assessment This problem assesses students' ability to describe geometric figures using words and/or diagrams, and to construct a triangle with given side lengths. If students have difficulty, you might advise them to begin by drawing a line segment.

15. Informal Assessment This problem assesses students' ability to use properties of triangles and to use construction when appropriate. You may want to have students make drawings to illustrate their answers.

Work Students Do

Students begin by tearing the angles off a triangle and putting them together on a straightedge to form a semicircle. Then students reverse the process, cutting a semicircle into three pieces that can be used to form the angles of a triangle. They discover that the sum of the angles of any triangle is 180°. Finally, students find the measurements of missing angles in triangles.

Goals

Students will:

- use properties of triangles (such as the sum of the angles, side relationships, and the Hinge theorem).

Pacing

- approximately three 45-minute class sessions

About the Mathematics

The sum of the angles in a triangle is 180°. In this section, students come to understand this property of triangles by constructing and deconstructing them. If the angles of a triangle are torn off and placed together on a straightedge with the angles next to each other, they form a semicircle with a "straight" angle of 180° (see illustration on page 40). Semicircles can also be torn into three angles that can make a triangle. Knowing that the sum of any triangle's angles is 180°, students can find missing angle measures by subtracting.

Materials

- drawing paper, pages 39, 41, 43, and 45 of the Teacher Guide (three or four sheets per student)
- centimeter rulers, pages 39, 41, 43, 45, and 47 of the Teacher Guide (one per student)
- scissors, pages 39, 41, 43, and 45 of the Teacher Guide (one pair per student)
- protractors or compass cards, pages 41, 43, 45, and 47 of the Teacher Guide (one per student)

Planning Instruction

Begin this section with the activity in which students tear the three angles off a triangle and arrange them along a straightedge. Students should repeat this activity several times so they can see that all triangles have angles that can be lined up along a straightedge.

Students may work on problems 5–7 and 12 in small groups. The remaining problems may be done individually.

There are no optional problems in this section.

Homework

Problems 10 and 11 (page 44 of the Teacher Guide) can be assigned as homework. After students complete Section C, you may assign appropriate activities from the Try This! section, located on pages 45–48 of the *Triangles and Beyond* Student Book. The Try This! activities reinforce the key mathematical concepts introduced in this section.

Planning Assessment

- Problems 13 and 14 can be used to informally assess students' ability to use properties of triangles (such as the sum of the angles, side relationships, and the Hinge theorem).

Activity

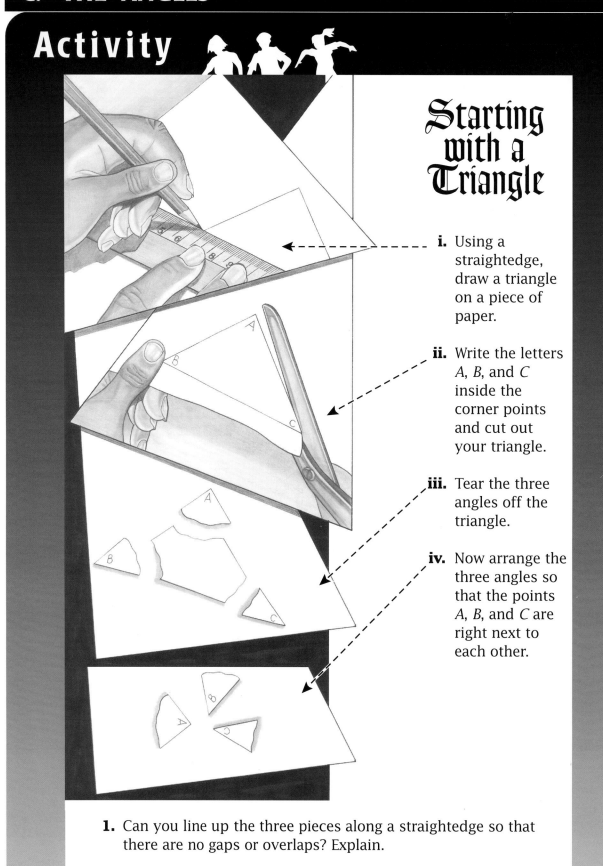

Starting with a Triangle

i. Using a straightedge, draw a triangle on a piece of paper.

ii. Write the letters A, B, and C inside the corner points and cut out your triangle.

iii. Tear the three angles off the triangle.

iv. Now arrange the three angles so that the points A, B, and C are right next to each other.

1. Can you line up the three pieces along a straightedge so that there are no gaps or overlaps? Explain.

1. Yes. The pieces may be lined up without gaps or overlaps on a straight line as shown below.

Materials drawing paper (one sheet per student); centimeter rulers (one per student); scissors (one pair per student)

Overview Students tear off the angles of a triangle and line them up along a straightedge.

Planning Students may do the activity and answer problem **1** individually. They may compare answers in small groups.

Comments about the Problems

1. This activity shows an important property of triangles: the sum of the angle measurements of a triangle is 180°. Although the activity is not a formal proof, it does help students to see that the angles together make a straight line.

Interdisciplinary Connection The musical instrument called a triangle consists of a metal triangle that is missing one of its angles. Like a drum, a triangle is a percussion instrument. It is made from a steel rod and hangs from a nylon loop. To play the triangle, a musician must strike it with another steel rod. One strike to a triangle can be heard even when an entire orchestra is playing.

Using another triangle, repeat the steps given on page 15:

- draw a triangle with a different shape;
- label each angle with a letter;
- cut out the triangle;
- tear off the angles;
- use a straightedge to line up the three angles.

2. Are there any gaps or overlaps this time?

3. Try to find a triangle whose angles will not line up along a straightedge.

4. Did you discover a geometric property of triangles from this activity? If so, what is it?

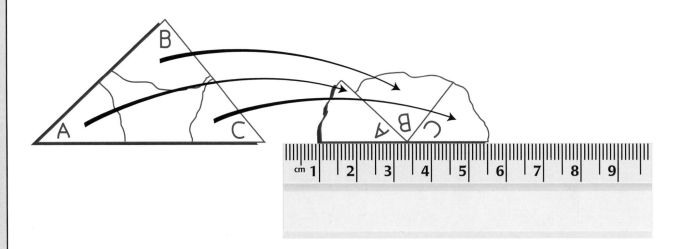

Shown above is a drawing that represents this geometric property. You can see that the three angles put together resemble a semicircle.

5. What is the sum of the angles in a semicircle?

6. Rewrite your geometric property using your answer from problem **5.**

2. No, there are no gaps or overlaps.

3. Such a triangle cannot be found. All triangles have angles that will line up along a straightedge.

4. Answers will vary. Sample responses:

- Every triangle has three angles that form a straight line when they are put together.

- When the three angles are put together, they form a semicircle.

5. 180°

6. The sum of the angle measurements of any triangle is 180°.

Materials drawing paper (one sheet per student); centimeter rulers (one per student); scissors (one pair per student); protractors or compass cards, optional (one per student)

Overview Students continue to make triangles, tear off the angles, and line up the angles along a straightedge. Then students generalize about their findings. They discover that the sum of the angles of a triangle is 180°.

Planning Students may work on problems **2–4** individually. Discuss problem **4** in class before students begin problems **5** and **6,** which can be done in small groups.

Comments about the Problems

2. Encourage students to try tearing the angles off in different ways to test this concept. This will help them understand what an *angle* is.

This activity can also be done without cutting, by folding a triangle as shown below:

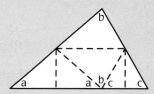

4. Problem **4** is critical because students are asked to formulate the main theorem of this section. If students have difficulty, you might suggest that they describe what they observed when they did the activity.

5. Some students may be able to answer this problem without measuring the angles. If students have difficulty, you might have them measure the angles with a protractor or compass card. You may want to review how to measure an angle. Students learned how to measure an angle with a compass card in the grade 5/6 unit *Figuring All the Angles.*

6. Students should rewrite the property in terms of the sum of the angles in degrees.

Activity

Starting with a Semicircle

i. Cut a semicircle out of a piece of paper (you don't have to be very precise).

ii. Then cut the semicircle into three pieces by making two cuts from the center of the straight side. Put the letters *A*, *B*, and *C* in the corner points.

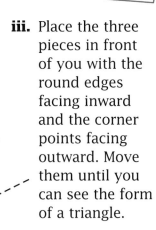

iii. Place the three pieces in front of you with the round edges facing inward and the corner points facing outward. Move them until you can see the form of a triangle.

iv. Now move the pieces a little further apart and closer together to make larger and smaller triangles.

Britannica Mathematics System

Students' semicircle pieces should look similar to the ones on page 17 of the Student Book.

Materials drawing paper (one sheet per student); centimeter rulers (one per student); scissors (one pair per student); protractors or compass cards (one per student)

Overview Students cut a semicircle into three pieces and form triangles using the pieces.

About the Mathematics This activity is the reverse of the previous activity. Now the semicircle is cut apart, and the pieces are used to form triangles. Students will describe this geometric property on the next page.

Planning Students may do this activity individually.

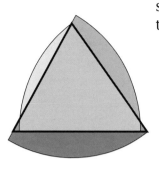

Repeat the steps on page 17 with another semicircle, and see if you can again make triangles.

7. Can you cut a semicircle into three pieces that will not form a triangle? Explain. (The three pieces should be formed by making two cuts from the center of the straight side.)

8. From this activity, you have discovered another geometric property. Write this property in your notebook.

Below you see a drawing of this geometric property. The three angles in the drawing add to 180°.

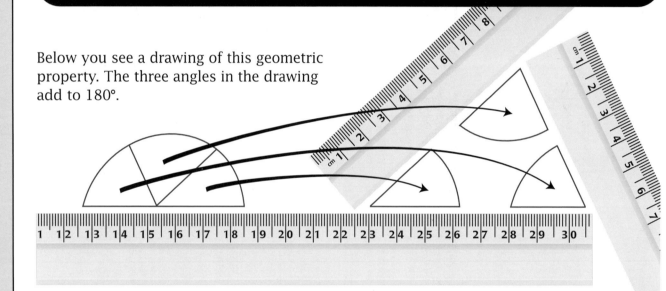

9. Rewrite your geometric property from problem **8** using this information.

By now you probably have several triangles that are separated into three pieces on your table.

10. Take three angles whose measures total more than 180° and try to make a triangle with them. Is this possible?

11. Try to make a triangle with three angles whose measures total less than 180°. Will this work?

7. No, any combination of angles cut this way from a semicircle will form a triangle.

8. Answers will vary. Sample response:

With every set of three angles that are cut from a semicircle, a triangle can be made.

9. Answers will vary. Sample response:

If you have three angles with measurements that add up to 180°, you can make a triangle from them.

10. No, a triangle cannot be formed.

11. No, a triangle cannot be formed.

Materials drawing paper (one sheet per student); centimeter rulers (one per student); scissors (one pair per student); protractors or compass cards (one per student)

Overview Students cut semicircles into three pieces and make triangles with the pieces. They generalize about their findings and formulate a rule that relates to any semicircle. They also try to find counterexamples.

About the Mathematics Any semicircle (or "straight" angle) can be cut into three pieces that will form a triangle.

Planning Students may work on problem **7** in small groups. Problems **8–11** may be done individually. Problems **10** and **11** may also be assigned as homework. You may want to discuss the activity in class after students have completed problem **9.**

Comments about the Problems

8. Like problem **4,** problem **8** is critical. Students are being asked to formulate the main theorem of this section in a different way. Again, this activity does not provide a formal proof, but it does help to make the idea more concrete for students. If students have difficulty, you may want to discuss problem **8** in class.

10–11. Homework These problems may be assigned as homework. They reinforce the concept that the sum of the angles in a triangle is 180°.

Students may want to use a compass card or protractor to create three angles that add to more (or less) than 180°, either by making three separate angles or by cutting apart an angle that is larger (or smaller) than 180°.

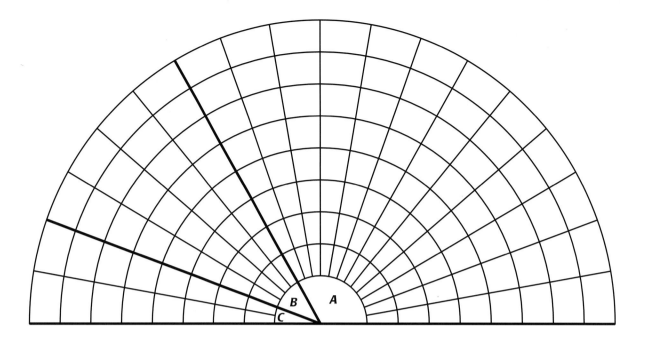

Above you see angles *A*, *B*, and *C* in a semicircle which has been subdivided into equal parts. They can be cut apart and put together to form triangle *ABC*.

12. **a.** What are the measures of angles *A*, *B*, and *C*?

    b. Side *AC* is 10 centimeters long. Draw side *AC* on a piece of paper.

    c. Angle *A* occurs at point *A*. Draw angle *A* on your piece of paper.

    d. Finish drawing triangle *ABC*.

    e. Is it necessary to use the measures of all three angles to finish drawing your triangle? Explain.

    f. There are many triangles with the same three angles as the triangle you drew in part **d,** but with different lengths for side *AC*. What can you tell about the shapes of these triangles?

12. a. Angle $A = 120°$

Angle $B = 40°$

Angle $C = 20°$

b.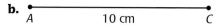

A   10 cm   C

c.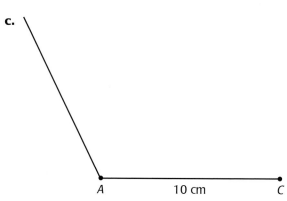

A   10 cm   C

d.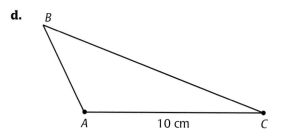

B

A   10 cm   C

e. No. The minimum information needed in this case is the length of side *AC* and the measurements of angle *A* and either angle *B* or angle *C*.

f. The triangles have the same shape, or are *similar,* and differ only in size.

Materials centimeter rulers (one per student); protractors or compass cards, (one per student)

Overview Students find the measurements for three angles. They construct a triangle using these measurements and the length of one side.

About the Mathematics At this point, you may introduce the idea of using three letters to identify a triangle. For instance, below is triangle *ABC*:

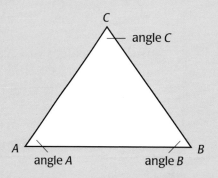

Planning Students may work on problem **12** in small groups. Discuss students' answers.

Comments about the Problems

12. a. Students may use protractors, or they may use the grid on Student Book page 19 to measure the angles. The grid is divided into 18 equal angles so that each section is 10°. Students should check to see whether the three angles add up to 180°.

   b. Students need to use a centimeter ruler to draw side *AC.*

   c. Encourage students to use protractors or compass cards to draw angle *A,* instead of tracing it.

Summary

The sum of the measures of the three angles of any triangle is 180°.

If the measures of three angles total 180°, then a triangle can be made with these angles.

If the measures of three angles do not total 180°, then a triangle cannot be made with these angles.

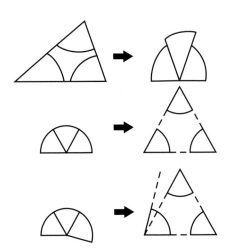

Summary Questions

13. In your notebook, sketch each of the following figures and fill in the value of the missing angle(s). (The drawings are not to scale, so do not try to measure them to find the answers.)

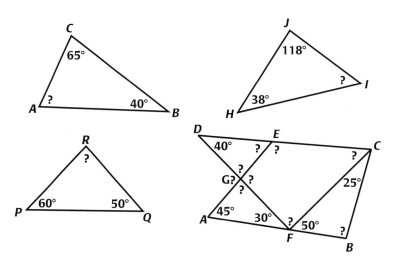

Imagine another triangle, *XYZ*, in which the measure of angle *Y* is twice that of angle *X*, and the measure of angle *Z* is three times that of angle *X*.

14. a. What is the measure of each angle?

b. Draw a triangle with these three angle measures.

13.

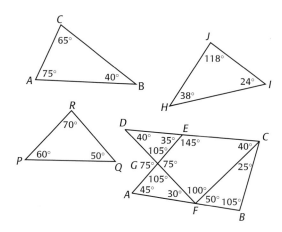

Planning Students may work on problems **13** and **14** individually. These problems may also be used as informal assessments. After students complete Section C, you may assign appropriate activities from the Try This! section, located on pages 45–48 of the *Triangles and Beyond* Student Book, for homework.

Comments about the Problems

13–14. Informal Assessment These problems assess students' ability to use properties of triangles (such as the sum of the angles, side relationships, and the Hinge theorem).

14. a. angle $X = 30°$

angle $Y = 60°$

angle $Z = 90°$

Students may use a variety of strategies to find the measurement of each angle. Sample strategy:

$X + Y + Z = 180°$
$Y = 2X$
$Z = 3X$
$X + Y + Z = X + 2X + 3X$
$6X = 180°$
So, $X = 30°$, $Y = 60°$, and $Z = 90°$.

13. Students may sketch the pictures and write the angle measurement next to each angle. You might also introduce a three-letter notation for angles. For example, the 35° angle at vertex *E* could be labeled angle *DEG*. Students could then list their answers next to the name of each angle.

Alternatively, students could label each angle using a number. Students should find the measurement of each angle by reasoning, not by measuring. Regardless of the method students use, they should explain how they found each answer.

14. Some students may already know that triangle *XYZ* is called a *right triangle*.

If students have difficulty, you might have them experiment by drawing triangles and labeling the angles in terms of *x* or using trial and error with the angle measurement. An algebraic solution (for instance, $x + 2x + 3x = 180$) may be too difficult for some students at this stage (See the solutions column.).

b.

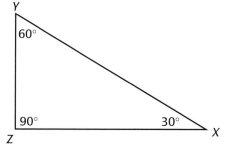

Work Students Do

Students begin this section by investigating which of two trees is closer to a given tree when it is impossible to measure distances directly. They consider each tree to be the vertex of a triangle, and measure the angle created by invisible lines going to the two other trees. Students use this information to determine which of the two trees is closer to the given tree by considering that the longest side of a triangle is opposite the largest angle. Next, students explore the relationship between sides and angles in a triangle in more detail with some actual triangles. They draw conclusions about isosceles and equilateral triangles based on their results. Students conclude this section by applying their newfound knowledge to find missing angles in given triangles without measuring them.

Goals

Students will:

- use properties of triangles (such as the sum of angles, side relationships, and the Hinge theorem);
- describe geometric figures using words and/or diagrams;
- create mathematical representations from visualization.

Pacing

- approximately two 45-minute class sessions

About the Mathematics

In a triangle, the largest side is opposite the largest angle, the second-largest side is opposite the second-largest angle, and the shortest side is opposite the smallest angle. This fact is known as the Hinge Theorem because the angle can be thought of as a hinge—the more it opens (the larger the angle) the larger the distance between the two sides (the longer the opposite side). The Hinge theorem can be used to establish that a triangle with equal base angles (isosceles) has two equal sides (and vice versa), and that a triangle with all angles equal (equilateral) has three equal sides (and vice versa).

The relationships between the sides and angles of a triangle are further developed in the unit *Looking at an Angle* and throughout the study of trigonometry.

Materials

- Student Activity Sheet 6, page 130 of the Teacher Guide (one per student)
- centimeter rulers, pages 57, 59, and 63 of the Teacher Guide (one per student)
- compass cards or protractors, pages 57 and 63 of the Teacher Guide (one per student)
- colored pencils, page 57 of the Teacher Guide (three per student)
- compasses, page 59 of the Teacher Guide (one per student)
- tracing paper, page 59 of the Teacher Guide (one sheet per student)
- scissors, page 59 of the Teacher Guide (one pair per student)
- pictures of triangles gathered by students on page 2 of the Student Book, page 63 of the Teacher Guide (several per student)

Planning Instruction

You might begin this section by asking students how they think that property boundaries are measured. Explain that surveyors, people who measure and mark points and boundaries, cannot always measure distances physically. Instead, they use geometry. Many surveyors' tools measure angles.

Students may work together as a class to solve problems 1 and 2. Problems 9–11 may be done individually. The remaining problems may be done individually or in small groups.

There are no optional problems in this section.

Homework

Problems 9 and 10 (page 60 of the Teacher Guide) can be assigned as homework. The Extension (page 63 of the Teacher Guide) may also be assigned as homework. After students complete Section D, you may assign appropriate activities from the Try This! section, located on pages 45–48 of the *Triangles and Beyond* Student Book. The Try This! activities reinforce the key mathematical concepts introduced in this section.

Planning Assessment

- Problem 5 can be used to informally assess students' ability to use properties of triangles (such as the sum of angles, side relationships, and the Hinge theorem) and their ability to create mathematical representations from visualization.
- Problem 9 can be used to informally assess students' ability to describe geometric figures using words and/or diagrams.
- Problem 11 can be used to informally assess students' ability to use properties of triangles (such as the sum of angles, side relationships, and the Hinge theorem).

For their math homework, Julia and Edgar have to find out which tree is closer to tree C: tree A or tree B. Unfortunately, tree C is across a stream from tree A and tree B.

1. a. Can you tell from the drawing which tree is closer to tree C?

   b. How can they find out which tree is closer?

1. a. You cannot tell which tree is closer from looking at the drawing. The two trees appear to be about the same distance from the third tree. In addition, it is possible that one tree is bigger than another, but appears the same size because it is farther away.

   b. Answers will vary. Students may suggest that they measure the distances with string.

Overview Students think about ways to decide which of two trees is closer to a third tree when it is impossible to pace off the distances.

Planning Students may work together as a class to solve problem **1.**

Comments about the Problems

1. a. Some students may think that they can see which tree is closer. Explain to students that there is no way to be sure just from looking at the drawing.

   b. Give students the opportunity to think of methods for determining which tree is closer without actually measuring the distances. Tell students that it would be difficult for Julia and Edgar to cross the stream.

Julia and Edgar do not want to cross the stream to find out which tree is closer to tree C.

Instead, Julia takes two meter sticks and a sheet of cardboard to tree A. She makes an angle with the two meter sticks, carefully lining one up with tree B and the other with tree C. She uses her cardboard to record the angle and labels it angle *A*.

Edgar takes two meter sticks and a sheet of cardboard to tree B. He makes an angle with his two meter sticks, lining them up with trees A and C. He uses his cardboard to record the angle and labels it angle *B*.

When Julia and Edgar compare their angles, they find that angle *A* is larger than angle *B*. Now they know which tree is closer to tree C.

2. Which tree do you think Julia and Edgar decide is closer to tree C? Why?

2. Tree A is closer to tree C.

Explanations will vary. Some students may try drawing different triangles to explore the relationship between the size of the angles and the length of the sides as shown below:

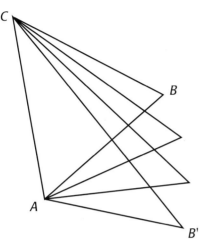

The distances from *A* to *C* and from *A* to *B* are fixed. As angle *A* gets larger, the distance from *B* to *C* also gets larger. So if angle *A* is larger than angle *B*, then *B* must be farther from point *C*.

Overview Students study the way Julia and Edgar solve the problem introduced on the previous page.

About the Mathematics The relationship between the size of an angle in a triangle and the length of the opposite side is explored in the rest of this section.

Planning Students may work together as a class to solve problem **2.**

Comments about the Problems

2. You might model the tree situation in your classroom or outside. For example, you could use desks to represent trees. It will be easier if the three model trees are positioned so that one of the angles (*A* or *B*) is obtuse, because it is easier to see the results if the differences in side lengths are large. If you do not model the problem, you may want to encourage students to draw a top view of the situation and label trees *A*, *B*, and *C*.

This problem is difficult, so it is likely that students will not be able to solve it at this point. If students have difficulty, simply encourage them to think about the problem and come up with arguments that support their thinking.

3. a. On **Student Activity Sheet 6,** measure the angles and sides of triangle *ABC* and label the drawing.

   b. Mark the triangle's largest angle.

   c. Mark the triangle's longest side.

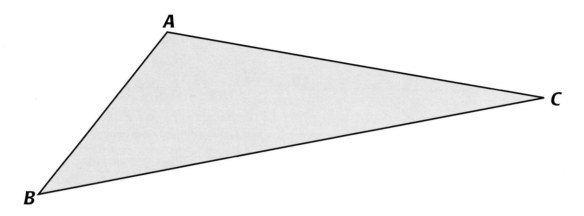

4. a. On **Student Activity Sheet 6,** with one color, mark the smallest angles and the shortest sides of triangles I, II, and III.

   b. Next, with a different color, mark the largest angles and the longest sides of triangles I, II, and III.

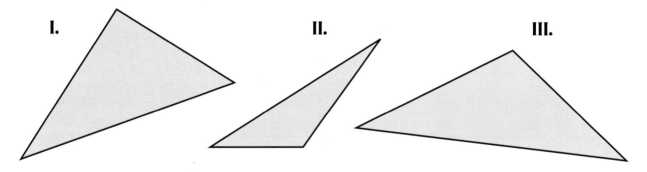

   c. Do you see another possible geometric property of triangles from the answers to problems **3** and **4?** If so, what is the property?

5. How do you think Julia and Edgar decided which tree is closer to tree C?

3. a–c.   angle *A*: 120°   side *AB*: 5.7 cm
         angle *B*: 39°   side *AC*: 10.5 cm
         angle *C*: 21°   side *BC*: 14.2 cm

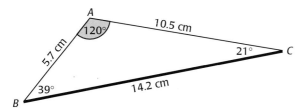

4.   a.

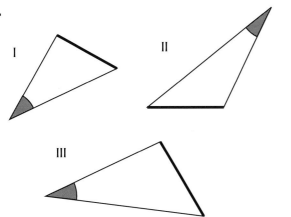

b.

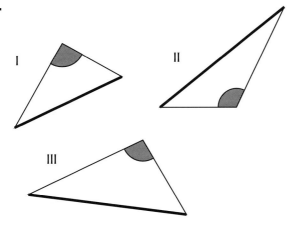

c. Answers will vary. Sample response:

The side opposite the largest angle is the longest; the side opposite the smallest angle is the shortest.

5. They used the relationship between the three angles and their opposite sides. Because angle *A* is larger than angle *B*, *BC* is longer than *AC*.

Materials Student Activity Sheet 6 (one per student); centimeter rulers (one per student); compass cards or protractors (one per student); colored pencils (three per student)

Overview Students label the longest and shortest sides and the largest and smallest angles of triangles. They discover that the longest side is opposite the largest angle and the shortest side is opposite the smallest angle. Using this information, students reconsider problem **2** (page 54 of the Teacher Guide).

Planning Students may work on problems **3–5** individually or in small groups. However, you may want to discuss problem **4c** in class. Problem **5** may be used as an informal assessment.

Comments about the Problems

3.   a. Students should measure the sides using a centimeter ruler. By now they should be able to use a compass card or protractor to measure angles without any problems.

b–c. Students may mark the sides on Student Activity Sheet 6 using colored pencils.

4. Students do not necessarily have to measure all sides and angles. For example, they may clearly see which side of triangle II is the longest and which angle is the largest.

5. Informal Assessment This problem assesses students' ability to use properties of triangles and their ability to create mathematical representations from visualization.

Using the property formulated in problem **4c**, students may finally solve problem **2** (page 54 of the Teacher Guide). Remind them that Julia and Edgar discovered that angle *A* is larger than angle *B*. Encourage students to draw a top view model of three trees and connect them with lines, forming a triangle. They can then investigate this triangle using the same methods they used to solve problems **3** and **4.**

Pictured below are two isosceles triangles. (Remember that isosceles triangles have at least two equal sides.) The slashes on the sides tell you that those sides are equal.

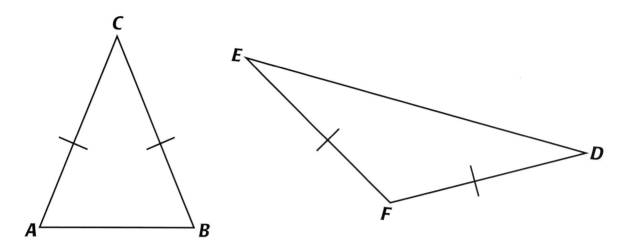

In triangle *ABC*, sides *AC* and *BC* have equal lengths, and side *AB* is the shortest side.

In triangle *DEF*, sides *DF* and *EF* have equal lengths, and side *DE* is the longest side.

6. a. Name the angles opposite sides *AC* and *BC*. What can you conclude about these two angles?

b. Name the angles opposite sides *DF* and *EF*. What can you conclude about these two angles?

7. Draw an isosceles triangle or use a compass to construct one. Cut out your triangle and fold it in half. What can you conclude about the angles of your isosceles triangle?

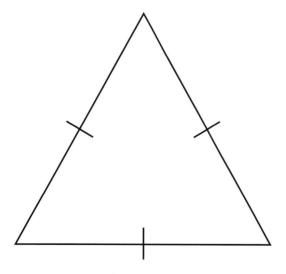

8. What can you conclude about the angles of an equilateral triangle such as the one shown on the left?

6. a. Angle *A* is opposite side *BC*. Angle *B* is opposite side *AC*. The measurements of angles *A* and *B* must be equal.

b. Angle *E* is opposite side *DF* and angle *D* is opposite side *EF*. The measurements of angles *E* and *D* must be equal.

7. By folding the triangle over the center line, you can see that the measurement of angle *A* equals the measurement of angle *B* as shown below.

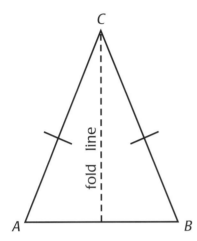

8. The three angles have equal measurements. Every triangle has angles with measurements that total 180°; 180° ÷ 3 = 60°. Therefore, every angle of an equilateral triangle must measure 60°.

Materials centimeter rulers (one per student); compasses (one per student); transparent paper (one sheet per student); scissors (one pair per student)

Overview Students use the relationship between angles and the sides opposite them to draw conclusions about the angles of isosceles and equilateral triangles.

About the Mathematics The Hinge Theorem can be used to establish that a triangle with equal base angles (isosceles) has two equal sides (and vice versa), and that a triangle with all angles equal (equilateral) has three equal sides (and vice versa).

Planning Students may work on problems **6–8** individually or in small groups. Discuss students' conclusions in class.

Comments about the Problems

6. Students may use one letter or three to name each angle. For example, angle *A* could also be named angle *CAB*.

7. If time is a concern, students can trace one of the isosceles triangles pictured at the top of the page instead of constructing one.

8. If students have difficulty, you might have them draw an equilateral triangle, cut it out, and fold it to compare its angles.

9. Copy and complete the following sentences describing the angles of isosceles and equilateral triangles.

> **In an isosceles triangle . . .**
>
> **In an equilateral triangle . . .**

10. In your notebook, sketch the following triangles and fill in the values of the missing angles without measuring them. (*Note:* The drawings are not to scale.)

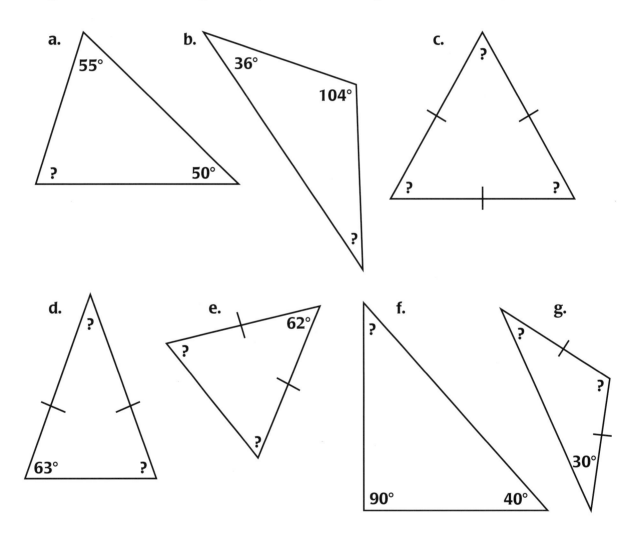

9. In an isosceles triangle, the angles opposite the two equal sides have the same measurement.

In an equilateral triangle, all the angles have the same measurement, 60°.

10. a–g.

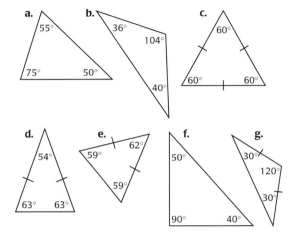

Overview Students complete sentences describing properties of the angles in isosceles and equilateral triangles. Then they solve problems using the properties of triangles they discovered in this section.

Planning Problems **9** and **10** may be done individually. They may also be assigned as homework. Problem **9** may also be used as informal assessment.

Comments about the Problems

9–10. Homework These problems may be assigned as homework.

9. Informal Assessment This problem assesses students' ability to describe geometric figures using words and/or diagrams.

You might want to have students compare equilateral and isosceles triangles. Ask them, *Are all equilateral triangles also isosceles triangles?* [yes]

10. Make sure that students can explain how they found their answers. Since students cannot measure the angles, they will have to find the values by reasoning. For example, the missing angle in problem **10a** can be found by subtracting the sum of the two given angles from 180°: 180° − (55° + 50°) = 180° − 105° = 75°. In problem **10c,** students may use the information that the three sides are equal to conclude that the three angles are equal. Since the three angles in every triangle add up to 180°, each angle measures 60°.

Students may conclude that since the triangle in problem **10g** is isosceles, one of the missing angles measures 30°, and the other must measure 120°; 180° − (30° + 30°) = 120°.

Summary

The largest angle of a triangle lies opposite the longest side.

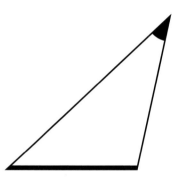

The second largest angle of a triangle lies opposite the second longest side.

The smallest angle of a triangle lies opposite the shortest side.

Isosceles triangles have two equal sides and two equal angles. Equilateral triangles have three equal sides and three equal angles.

Summary Questions

11. a. Choose any triangle from among the pictures that you collected at the beginning of the unit. Label the largest angle and the smallest angle. Justify your answer using only a ruler.

b. Choose another triangle, and label the longest side and the shortest side. Justify your answer using only a compass card or protractor.

11. a–b. Answers will vary, depending on students' triangles. However, the longest side is always opposite the largest angle, and the shortest side is always opposite the smallest angle.

Materials centimeter rulers (one per student); compass cards or protractors (one per student); pictures of triangles gathered by students on page 2 of the Student Book (several per student)

Overview Students read the Summary, which reviews the main concepts covered in this section. Then students label the largest and smallest sides and angles of triangles in their collection.

Planning Students may work on problem **11** individually. It may also be used as informal assessment. Discuss students' answers in class, using one or two examples. After students complete Section D, you may assign appropriate activities from the Try This! section, located on pages 45–48 of the *Triangles and Beyond* Student Book, for homework.

Comments about the Problems

11. Informal Assessment This problem assesses students' ability to use properties of triangles. Some students may not have pictures of triangles in their collections that are easy to work with. In that case, you might ask students to describe how to find the largest and smallest angles of a triangle using only a ruler, and how to find the longest and shortest sides of a triangle using only a protractor or compass card.

a. Students may not need a ruler to find the longest sides of some triangles. If they know which side is the longest, they should also know which angle is the largest.

b. Students should use compass cards to determine which angle is the largest and the smallest.

Extension The theorem students discover in this section is called the Hinge theorem. This theorem states that in a triangle, the largest side is opposite the largest angle, the second largest side is opposite the second largest angle, and the smallest side is opposite the smallest angle. Ask students, *Why do you think this rule is called the Hinge theorem?* [Each angle in a triangle can be thought of as a hinge. The more it opens (the larger the angle), the larger the distance between the two sides that form the angle. As a result, the wider the angle, the longer the side opposite the angle.]

Work Students Do

Students examine a photograph of a harvested field and identify what is special about the lines formed by the rows of harvested crops. They describe the difference between lines that are parallel and lines that are not parallel. Students learn how to draw parallel lines using a triangle and a straightedge. They also investigate some situations in which understanding the properties of parallel lines is necessary, including an optical illusion and an altered perspective of parallel train tracks. Students then construct parallelograms using paper strips that represent pairs of parallel lines, and learn to identify different types of parallelograms.

Goals

Students will:

- use properties of triangles (such as the sum of the angles, side relationships, and the Hinge theorem);*
- use construction when appropriate;
- describe geometric figures using words and/or diagrams;
- recognize the need for mathematical rigor in justifying answers.

* This goal is assessed in other sections of the unit.

Pacing

- approximately three 45-minute class sessions

Vocabulary

- family of parallel lines
- parallel
- parallelogram
- quadrilateral
- rhombus

About the Mathematics

This section introduces the properties of parallel lines: (1) they are always the same distance apart; (2) they never touch; (3) they form equal angles with a line that crosses them (a transversal).

The properties of special parallelograms are also explored in this section. A rectangle is a parallelogram with four equal angles that each measure 90°. A rhombus is a parallelogram with four equal sides, but not necessarily four equal angles. A square is a parallelogram with four equal sides and four equal angles. In any parallelogram, opposite angles are congruent.

Materials

- Student Activity Sheet 7, page 131 of the Teacher Guide (one per student)
- compass cards or protractors, page 69 of the Teacher Guide (one per student)
- transparent paper, page 69 of the Teacher Guide, optional (one sheet per student)
- straightedges, pages 71, 73, 75, 77, and 79 of the Teacher Guide (one per student or group of students)
- plastic or cardboard triangles, pages 71, 73, 75, and 79 of the Teacher Guide (one per student)
- compasses, pages 73 and 75 of the Teacher Guide, optional (one per student)
- colored pencils, page 73 of the Teacher Guide, optional (one box per student)
- pictures or drawings of buildings from newspapers or magazines, page 75 of the Teacher Guide, optional (several per student)
- drawing paper, page 77 of the Teacher Guide (one sheet per group of students)
- transparencies, pages 77 and 79 of the Teacher Guide (one sheet per student or group of students)
- scissors, page 79 of the Teacher Guide (one pair per student or group of students)
- colored markers, page 81 of the Teacher Guide, optional (one box per group of students)
- colored drinking straws, page 81 of the Teacher Guide, optional (one box per group of students)
- colored paper, page 81 of the Teacher Guide, optional (several sheets per group of students)

Planning Instruction

You may want to begin this section with a discussion of the photograph on page 27 of the Student Book. Ask students if they have seen lines that are similar to these anywhere else. If students have difficulty thinking of any examples, you might have them look for parallel lines in the classroom.

Students may work on problems 1, 7, 15, and 16 individually. Problems 13 and 14 may be done individually or in small groups. The remaining problems may be done in small groups.

Problem 9 is optional. If time is a concern, you may omit this problem or assign it as homework.

Homework

Problems 7 (page 72 of the Teacher Guide) and 13 and 14 (page 78 of the Teacher Guide) may be assigned as homework. The Extensions (pages 69, 75, and 81 of the Teacher Guide) and the Writing Opportunities (pages 69, 77, and 79 of the Teacher Guide) may also be assigned as homework. After students finish Section E, you may assign appropriate activities from the Try This! section, located on pages 45–48 of the *Triangles and Beyond* Student Book. The Try This! activities reinforce the key mathematical concepts introduced in this section.

Planning Assessment

- Problem 4 can be used to informally assess students' ability to use construction when appropriate and to recognize the need for mathematical rigor in justifying answers.
- Problems 15 and 16 can be used to informally assess students' ability to describe geometric figures using words and/or diagrams.

E. PARALLEL

SIDE BY SIDE

1. What is special about the lines in this photograph?

1. Answers will vary, but these lines are (roughly) parallel. Sample student responses:

   They are straight lines.

   They run alongside each other.

   The lines never touch.

   The lines all run in the same direction.

Overview Students look at a photograph of parallel rows of crops in a field. They consider the properties of the lines formed by the crops.

About the Mathematics Parallel lines in a two-dimensional picture may have a vanishing point. Students should realize that although the lines seem to come closer to each other in the picture, they are parallel in reality.

Planning Students may work on problem **1** individually. You may want to have another class discussion after students have answered the problem.

Comments about the Problems

1. Allow students to use their own words to describe the lines in the photograph. Some students may notice that the field is harvested and that the distance between the rows is the same. Other students may think that the rows get closer to each other at the end of the field. If so, explain that the rows only appear to get closer.

Here is a diagram of an aerial view of a different field.

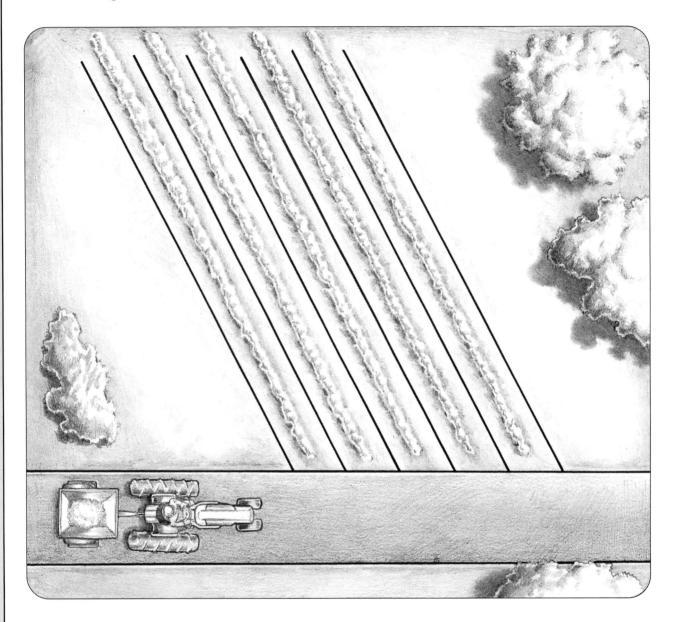

The lines in the above field are parallel. The word parallel comes from a Greek word meaning "side by side."

2. Select two parallel lines in the above diagram.

  a. Describe the distance between these two lines at several places.

  b. Measure the angles that these two lines make with the road. What can you conclude?

3. Draw two lines that are not parallel. Describe two ways that you can recognize lines that are not parallel.

2. Answers will vary. Sample responses:

   a. They are an equal distance apart everywhere (1.4 cm).

   b. The angles they make with the road are equal (about 118° measured from the right-hand side).

3. Drawings will vary. Sample drawings:

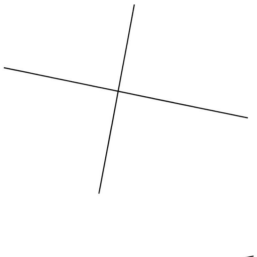

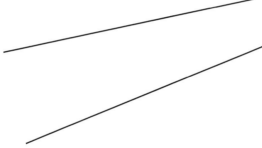

Descriptions will vary. Sample responses:

Two lines are not parallel if they cross (or if they would cross when extended far enough).

Two lines are not parallel if they are not always the same distance apart.

Two lines are not parallel if the angles they make with a third line that intersects both of them are not equal.

Materials compass cards or protractors (one per student); transparent paper, optional (one sheet per student)

Overview Students investigate some characteristics of parallel lines.

About the Mathematics The properties of two parallel lines are as follows:

• The distance between two parallel lines is always the same.

• Two parallel lines never intersect.

• If two parallel lines are crossed by a third line, the two parallel lines make equal angles with that line.

Planning Students may work on problems **2** and **3** in small groups. You may want to discuss problem **2** with the whole class.

Comments about the Problems

2. b. Students may measure the angles with a compass card or protractor and compare the measurements. Another way to compare the angles is to use transparent paper. Students can trace one angle and put it on the other angles in order to see whether it fits. When students discover that the two angles are equal, ask them, *Do parallel lines always make equal angles with a line that crosses them?* [yes]

Extension You might want to challenge students to trace two of the parallel lines from the picture. Then they should draw another line that is parallel to the two lines they traced. Explain to students that there are an infinite number of places where this third line can be drawn. The third line can even be drawn between the two parallel lines and still be parallel to them.

Writing Opportunity You may ask students to write their answers to problem **3** in their journals.

Activity

Now you will learn how to draw parallel lines like a draftsperson or a designer.

i. Using a straightedge, draw a straight line. Place a plastic or cardboard triangle alongside the line (a triangle of any shape will work).

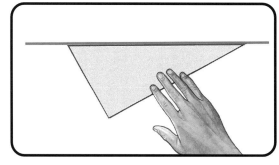

ii. Place the straightedge against another side of the triangle as shown.

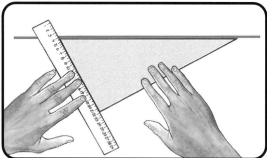

iii. While holding the straightedge still, slide the triangle along the straightedge.

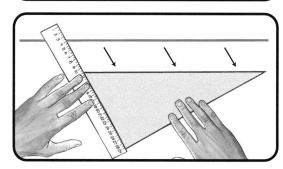

iv. Draw a second line along the edge of your triangle that is parallel to the first line.

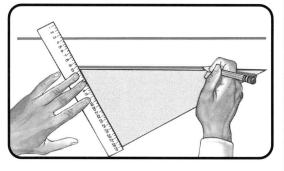

4. Why can you be sure that these two lines are parallel?

4. The two lines are parallel because the angles they make with the straightedge are equal (because it's the same angle in the triangle).

Materials straightedges (one per student); plastic or cardboard triangles (one per student)

Overview Students learn how to draw parallel lines.

About the Mathematics This activity involves the converse of the previous page; that is, if lines are constructed so that they form equal angles with a line intersecting them, the lines are parallel. Note that this is only true for lines in a two-dimensional situation, or in other words, if the lines are in the same plane. (In a three-dimensional situation, it is possible that two lines make the same angle with a third line without being parallel.)

Planning Students may work on the activity and answer problem **4** in small groups. Problem **4** can also be used as an informal assessment.

Comments about the Problems

4. **Informal Assessment** This problem assesses students' ability to use construction when appropriate and to recognize the need for mathematical rigor in justifying answers.

   Some students may respond that they know the lines are parallel because the lines do not cross. If so, ask students, *How can you be sure that the lines will never cross?* [They cannot cross because they were both drawn at the same angle to the straightedge.] This question can be used to guide students toward the idea of rigorous justification, which is an important part of mathematical reasoning. If students have difficulty, you might have them reconsider the answer to problem **2b.** Both these problems link parallelism to the concept of equal angles.

Activity

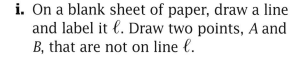

i. On a blank sheet of paper, draw a line and label it ℓ. Draw two points, *A* and *B*, that are not on line ℓ.

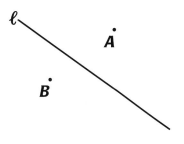

ii. Use a straightedge and your triangle from the activity on page 29 to draw a line through *A* that is parallel to line ℓ, and label it *m*.

iii. Use your straightedge and triangle to draw a line through *B* that is parallel to line ℓ, and label it *n*.

5. Are lines *m* and *n* parallel? If they are, you can write $m \parallel n$. The symbol $\parallel$ means "is parallel to."

6. Are the seven long lines in the box on the right parallel? Explain.

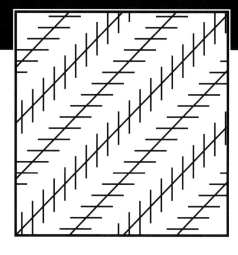

Look at the railroad tracks pictured on the left.

7. a. Make a top-view drawing of the six rails.

    b. In your drawing, mark the sections of the six rails that have the same lengths. Also mark the angles that have the same measures. (You should not use a ruler or protractor to determine the measures.)

5. Yes, the lines are parallel, as shown below.

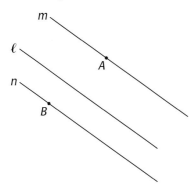

6. Yes, the lines are parallel. There are several ways to check. One way is to align one edge of a triangle on a line and place a straightedge along another side of the triangle. Then, slide the triangle along the straightedge. If the edge of the triangle aligns with each of the lines in the picture, the lines are parallel.

Another way to see whether the lines are parallel is to use a piece of paper. Line up the paper on one line, make a mark at a second line, and slide the paper along the line as shown below. If the lines are always the same distance apart, then they are parallel.

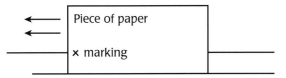

You can also check parallelism with a compass. Adjust the compass until the center just touches one line and the radius just touches another line. Draw several semicircles from different points on the original line. If they all just touch the second line, you can conclude that the lines are parallel.

Another method is to extend one side of the square border and extend some of the seven lines so that they all intersect the extended border. Then measure the angles formed by the seven lines and one side of the square border. If the angles are the same, then the lines are parallel.

7. a–b.

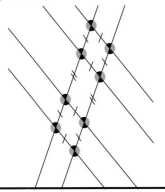

Materials straightedges (one per student); plastic or cardboard triangles (one per student); compasses, optional (one per student); colored pencils, optional (one box per student)

Overview Students continue to draw parallel lines. They investigate an optical illusion to determine whether lines are parallel. Finally, students make a top-view drawing of six rails and label equal line lengths and equal angles.

Planning Students may work on the activity and problems **5** and **6** in small groups. They may work on problem **7** individually. If time is a concern, problem **7** can also be assigned as homework. However, be sure to discuss students' answers in class.

Comments about the Problems

5. If students have difficulties with the activity, you may ask them first to draw two parallel lines. Then, ask students which instrument they moved and which they held in place. You may need to explain to students that in the activity they have to put one side of the triangle along line ℓ, place the straightedge along another side of the triangle, and then slide the triangle along the edge of the straightedge until they reach point *A*.

6. See whether students can explain how to find out whether the lines are parallel before actually trying it. You may want to have compasses available in case students want to use them to check the lines. Explain that it is possible to test for parallelism visually by holding the bottom left corner of the page up close to your eye. Viewed from this angle, the lines look parallel.

7. Homework This problem may be assigned as homework. You may want to have students use colored pencils to indicate equal parts of the rails and equal angles. Note that the lines do not have to intersect at right angles. Students should explain how they know which lines are equal and which angles are equal. The lines shown in the solutions column are equal because parallel lines are always the same distance apart. The angles are equal because parallel lines make equal angles with any line intersecting them.

8. This truck is moving down a ramp into a tunnel. On **Student Activity Sheet 7,** locate the point at which the truck will be completely hidden from view by the wall. Show how you located this point.

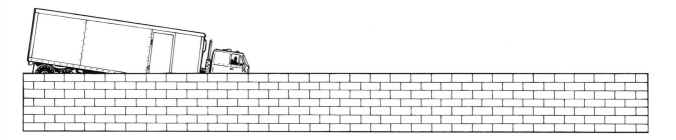

The above picture is of the National Aquarium in Baltimore, Maryland. This building has a very unique roof structure. Within its triangular faces, there are several "families" of parallel lines. A *family of parallel lines* is a set of lines that are all parallel to each other.

9. a. Choose one triangular face shown above. How many families of parallel lines do you see on that face?

   b. Make a drawing of that face and highlight each family of parallel lines with a different color.

8. Students should draw a line from the top of the truck to the ramp wall as shown below. The truck will be hidden from view at the point where the line touches the wall.

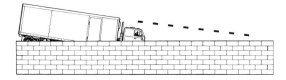

The assumption is that the road continues at the same level of steepness. The road is parallel to the roofline of the truck drawn.

9. Answers will vary depending on the triangular face students choose. Sample response:

   a. There are two families of parallel lines on the side facing the reader: one family of lines runs vertically, and the other family of lines runs horizontally.

   b.

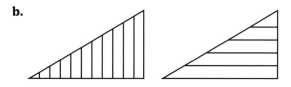

Materials Student Activity Sheet 7 (one per student); straightedges (one per student); plastic or cardboard triangles (one per student); colored pencils, optional (one box per group); pictures or drawings of buildings from newspapers or magazines, optional (several per student)

Overview Students solve a problem that can be explained using parallel lines. Then they investigate and draw families of parallel lines in a picture.

Planning Students may work on problems **8** and **9** in small groups. Problem **9** is optional. If time is a concern, it may be omitted or assigned as homework.

Comments about the Problems

8. If students have difficulties, it may be helpful to ask, *What is the last part of the truck you can see just before it disappears?* [the upper left corner] Students should see that parallel lines can be used to solve the problem. Students may see parallel lines at the roof and the bottom of the truck. They should also realize that if the road were shown in this picture, it would be a line parallel to the bottom and top of the truck. It might be interesting to ask students how they could solve the problem if the truck had a curved roofline.

Extension You may want to ask students to look for pictures or drawings of buildings and roads in newspapers. Then, ask students to indicate parallel lines or families of parallel lines in their pictures. You may display their results in your classroom.

Activity

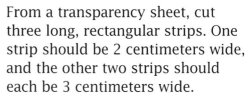

Parallelograms

From a transparency sheet, cut three long, rectangular strips. One strip should be 2 centimeters wide, and the other two strips should each be 3 centimeters wide.

Place two strips across each other as shown. Where the strips overlap, you will see a special *quadrilateral* (four-sided figure) called a parallelogram.

10. Can you tell what is special about this shape from the name parallelogram?

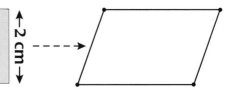

You can move the transparencies around to make different parallelograms with different shapes.

Use the 2-centimeter strip and one 3-centimeter strip to draw one set of parallelograms. You can mark the corners with a pencil and use your straightedge to draw the sides. Then draw another set of parallelograms using the two 3-centimeter strips.

11. a. List some of the similarities and differences between each set of parallelograms.

b. In your notebook, write some characteristics of a parallelogram. How are parallelograms different from other quadrilaterals?

10. Answers will vary. Sample student response:

The opposite sides are parallel.

11. a. Answers will vary. Sample responses:

The sides of each parallelogram made with the two 3-centimeter strips are always equal and are at least 3 centimeters long. (They are exactly 3 centimeters long if the parallelogram is a square.)

Each parallelogram made with one 2-centimeter strip and one 3-centimeter strip has two sides that are at least 2 centimeters long and two sides that are at least 3 centimeters long (some may be rectangles).

All the parallelograms have opposite angles equal, opposite sides of equal length, and opposite sides parallel.

b. Answers will vary. Sample response:

A parallelogram has four sides. Opposite sides have the same length and are parallel. Also, opposite angles are equal. Other quadrilaterals, like the one shown below, may have four sides that are of different lengths that are not parallel, and four angles that are not equal.

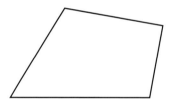

Materials straightedges (one per group of students); rectangular transparency strips, 2 centimeters wide (one per group of students); rectangular transparency strips, 3 centimeters wide (two per group of students); drawing paper (one sheet per group of students)

Overview Students use transparency strips to construct parallelograms. Then they look for characteristics of a parallelogram.

About the Mathematics On this page, a parallelogram is presented as a special quadrilateral with parallel sides. On the next page, rectangles, rhombi, and squares are introduced as special parallelograms.

Planning Students may work on problems **10** and **11** in small groups. You may wish to discuss students' answers to problem **10** before they answer problem **11**.

Comments about the Problems

10. Note that the only real requirement for the strips is that they have parallel edges. Parallelograms can also be constructed with two rectangular strips of paper. Strips of transparencies make it easier to visualize the parallelogram, however.

Writing Opportunity You may ask students to write their answers to problem **11b** in their journals.

12. Decide whether or not the following figures are parallelograms and justify your answers.

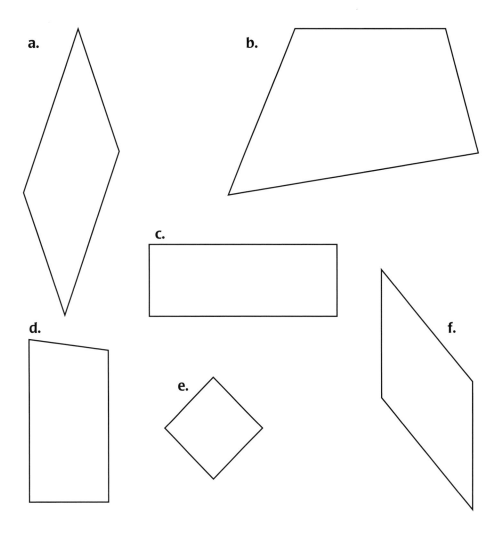

a.

b.

c.

d.

e.

f.

Some parallelograms have special shapes and special names.

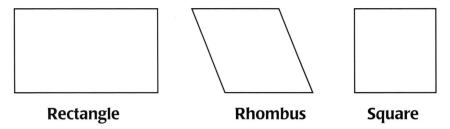

Rectangle   **Rhombus**   **Square**

13. Describe the characteristics that make each of the above parallelograms special.

14. Which of the parallelograms that you made for problem **11** on page 32 look similar to the special ones above?

12. Figures **a, c, e,** and **f** are parallelograms; **b** and **d** are not. Students may justify their answers by measuring the opposite sides or the opposite angles of each parallelogram, by using a straightedge and triangle to check that opposite sides are parallel, or by making new strips that match the measurements of the parallelograms.

13. Answers will vary. Sample response:

A rectangle is a parallelogram with four equal angles that each measure 90°. Most rectangles have one pair of sides that is longer than the other pair.

A rhombus is a parallelogram with four equal sides, but not necessarily four equal angles.

A square is a parallelogram with four equal sides and four equal angles. A square can also be defined as a rectangle and a rhombus.

14. Answers will vary. Students may note that all parallelograms made with the pair of 3-centimeter strips are rhombuses or squares. None of those made with the 2-centimeter strip and the 3-centimeter strip together are rhombuses or squares, but some may be rectangles.

Materials straightedges, optional (one per student); plastic or cardboard triangles, optional (one per student); transparencies, optional (one per student or group of students); scissors, optional (one pair per student or group of students)

Overview Students investigate different quadrilaterals and decide whether or not each figure is a parallelogram. Then they list characteristics of a rectangle, a rhombus, and a square.

About the Mathematics Equal sets of opposite angles or equal sets of opposite sides are each sufficient criteria for identifying parallelograms.

Planning Students may work on problems **12–14** individually or in small groups.

Comments about the Problems

12. You may want to have triangles and straightedges available for students who want to be certain of their answers. Some students may find pairs of parallel lines either by looking at them or by actually measuring them. Students cannot use the transparency strips from page 32 of the Student Book, because those strips are not the correct width. Students will have to make new strips.

13–14. Homework These problems may be assigned as homework.

13. Note that all of these figures are parallelograms.

14. You may want to ask students what a rhombus and a square have in common and how they differ.

Writing Opportunity You may ask students to write their answers to problem **13** in their journals.

Summary

Parallel lines do not intersect; they are always the same distance apart. Parallel lines form equal angles with lines that cross them.

A *parallelogram* is a four-sided figure formed by the intersection of two pairs of parallel lines. Rectangles, rhombuses, and squares are special kinds of parallelograms.

Summary Questions

15. Describe how you can determine if a shape is a parallelogram.

16. If a shape is a parallelogram, how can you tell if it is one of the following:

    a. a rectangle

    b. a rhombus

    c. a square

15. Descriptions will vary. Sample description:

A shape is a parallelogram if it is a quadrilateral and its opposite sides are parallel. So, I could measure the angles or measure the distances between opposite sides.

16. a. A rectangle has four right angles.

b. A rhombus has four equal sides.

c. A square has both four right angles and four equal sides.

Materials colored markers, optional (one box per group); colored drinking straws, optional (several per group); colored paper, optional (several sheets per group)

Overview Students read the Summary, which reviews the main concepts of this section. Then they solve two more problems about parallelograms.

Planning Before students begin the problems, you may want to discuss the Summary in class. Ask students to restate each statement in their own words and explain how they can check each statement. Students may work on problems **15** and **16** individually. These problems may also be used as informal assessment. After students finish Section E, you may assign appropriate activities from the Try This! section, located on pages 45–48 of the *Triangles and Beyond* Student Book, for homework.

Comments about the Problems

15–16. Informal Assessment These problems assess students' ability to describe geometric figures using words such as *parallel, parallelogram, rhombus, rectangle,* and *square.*

Extension You may want to invite groups of students to design a poster about the four quadrilaterals: the parallelogram, the rectangle, the rhombus, and the square. Students will have to decide how they can show the characteristics of each. You may wish to have various materials available for students to use, such as colored markers, colored drinking straws, and colored paper.

Work Students Do

Students learn how shapes can be flipped, turned, or slid to create a desired effect. They use a stencil to make a variety of new images from the original. Students learn how reflecting a shape across a line affects the shape. Then they are introduced to the terms *reflection, rotation,* and *translation* and shown how these transformations can be used to create new shapes. Students look for symmetry in pictures, letters, and shapes.

Goals

Students will:

- understand and identify line symmetry;
- describe transformations (translation, rotation, and reflection) using words and/or diagrams.

Pacing

- approximately two 45-minute class sessions

Vocabulary

- congruent figures
- line of reflection
- line of symmetry
- line symmetry
- reflection
- rotation
- translation

About the Mathematics

Reflection, rotation, and translation are three ways of manipulating a plane figure; they correspond to "flips," "turns," and "slides."

A reflection is the result of flipping a figure over a line (the line of reflection). It is important to realize that an even number of reflections will produce the original figure, and an odd number of reflections will produce a mirror image of the original figure. In addition, reflecting a symmetric figure over a line of symmetry will always give the original figure.

A translation is the result of sliding a figure.

A rotation is the result of turning a figure around a particular point. Rotation angles are measured using that point as a vertex and lines that touch the same spot in both figures (see illustration on the right).

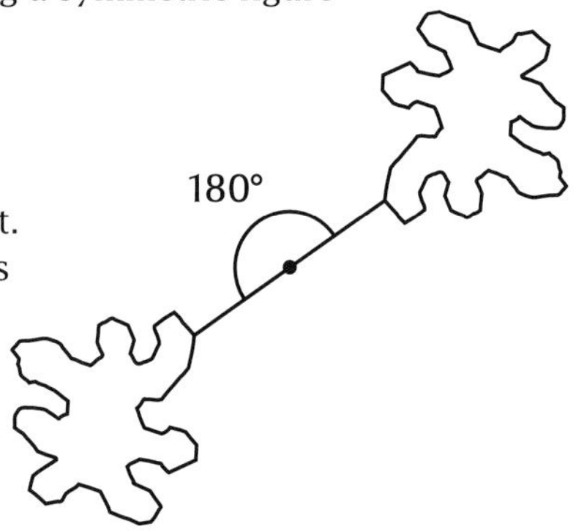

180°

Materials

- drawing paper, pages 87 and 91 of the Teacher Guide (one sheet per student)
- scissors, pages 87 and 91 of the Teacher Guide (one pair per student)
- congruent triangles, page 89 of the Teacher Guide, optional (two per class)
- overhead projector, page 89 of the Teacher Guide, optional (one per class)
- hand-held mirror, page 93 of the Teacher Guide, optional (one per class)

Planning Instruction

You may wish to begin the section with a discussion about stencils. You could ask students, *Why are stencils useful for making designs?* [Because they make consistent shapes, which when used a certain way, make other consistent shapes.] You may want to make a stencil in advance and demonstrate on the blackboard or with an overhead projector the design that can be made by turning, flipping, and sliding the stencil. You also might want to invite students to take their turns in creating the design in this way.

Students may work on problems 2, 3, and 11–13 in small groups. They may work on the remaining problems individually.

Problem 7 is optional and can be omitted or assigned as homework if time is a concern.

Homework

Problem 5 (page 88 of the Teacher Guide) and problem 6 (page 90 of the Teacher Guide) may be assigned as homework. The Extension (page 85 of the Teacher Guide) may also be assigned as homework. After students complete Section F, you may assign appropriate activities from the Try This! section, located on pages 45–48 of the *Triangles and Beyond* Student Book. The Try This! activities reinforce the key mathematical concepts introduced in this section.

Planning Assessment

- Problems 5 and 14 may be used to informally assess students' ability to describe transformations (translation, rotation, and reflection) using words and/or diagrams.
- Problems 9 and 15 may be used to informally assess students' ability to understand and identify line symmetry.

STAMPS AND STENCILS

Mark buys a set of monkey stamps at a toy store for his younger sister Mia. Mia uses her stamp set to make the above design.

1. **a.** What is the minimum number of stamps Mia can use to make the design? Explain your answer.

    b. The toy store also sells animal stencils such as the one on the left. If Mia purchases stencils instead of stamps, how many stencils does she need to make her design?

    c. Explain how to use the stencil on the left to draw each of the above monkeys.

1. a. Mia needs two stamps because there are two different monkeys, one with the tail pointing left and one with the tail pointing right. The monkeys are reflections of each other. Because a stamp cannot be *flipped,* she needs two different stamps.

b. She needs only one stencil because a stencil can be flipped, or turned over, to create a reflection of the original monkey.

c. Answers may vary. Sample response:

Monkey **A** can be drawn using this side of the stencil.

Monkey **B** is a rotation of **A,** so this side of the stencil still works.

Monkey **C** requires a flip of the stencil and a rotation.

Monkey **D** is a flip of **C,** so we must go back to the original side of the stencil.

Monkey **E** is a flip of **D,** so we must use the other side of the stencil.

In short, **A, B,** and **D** can be made with one side of the stencil, and **C** and **E** with the other side.

Overview Students investigate making pictures using a stamp and a stencil. They discover that a stencil has the advantage that it can be flipped over to create a reflection. In this context, students start to see how flips, turns, and slides affect pictures.

About the Mathematics This section informally introduces the three transformations: reflection (flip over), rotation (turn), and translation (slide).

Planning When students finish problem **1,** you may want to discuss the answers with them. Students may work on this problem individually.

Comments about the Problems

1. It is important that students use their own words in answering this problem. Formal terminology will be introduced on Student Book page 37. Students' answers should mention all three transformations: flips, slides, and turns.

  c. Students may describe all possibilities in terms of monkey **A**. If students say only which side of the stencil is used, encourage them to make their descriptions more precise.

Extension You may ask students to bring their own stencils to class. Then they may investigate these stencils to see whether or not the picture is different when the stencil is flipped. (For some stencils, it may not make a difference if it is flipped.)

Activity

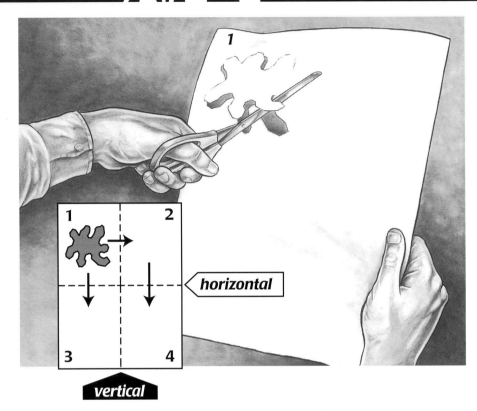

i. Cut out an irregular shape in the upper left corner of a sheet of paper. Put the number 1 in the corner near the cutout.

ii. Fold the paper in half vertically. Then use the hole as a stencil and trace the shape on the right side of the paper.

iii. Unfold the paper and cut out your new shape. Put the number 2 near the new cutout.

2. What is the relationship between the two cutouts?

iv. Fold your paper in half horizontally, and trace shapes 1 and 2 onto the bottom of the page. Put the number 3 in the lower left corner and the number 4 in the lower right corner.

3. a. What is the relationship between shapes 2 and 3?

   b. What is the relationship between shapes 1 and 4? Why do you think this happened?

2. The two cutouts are mirror images of each other. The second shape is the first one flipped over the vertical line.

3. a. The third shape is the second shape turned upside down (rotated 180°).

   b. The fourth shape is the first shape turned upside down (rotated 180°).

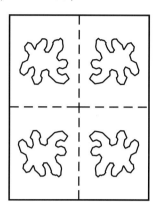

Materials drawing paper (one sheet per student); scissors (one pair per student)

Overview Students make a stencil and use this stencil to make a variety of new images by folding and tracing the original shape. Then they look for relationships between the shapes they have made.

Planning Have students work on the activity and problems **2** and **3** in small groups. Have a class discussion about students' answers when students have finished problem **3.**

Comments about the Problems

   2–3. Encourage students to use expressions such as *flip, turn,* and *slide.*

   2. It is important that students cut out an irregular shape (the shape of a hand works well); otherwise they may not see the different transformations. You may wish to use a mirror to show why certain images are called *reflections,* or *mirror images.*

   3. If the folds are perpendicular, this will work. If the folds are not perpendicular, the images cannot be superimposed exactly.

     a. The two shapes are rotations (turns) of each other.

     b. The two shapes are rotations of each other. Two flips give the original figure (rotated, depending on the angle of the flip).

Figures that are copies of each other are called *congruent figures.* Congruent figures have the same size and shape; they can fit on top of each other exactly with no overlaps.

4. Are triangles A and B congruent? Explain.

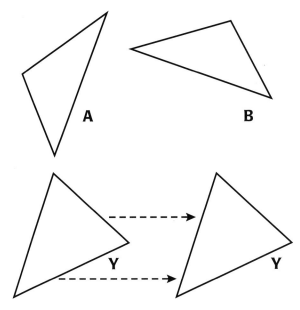

Terms

You can move figures around to see if they are congruent. These movements have three special names.

Translation—sliding a figure so that each point moves the same distance in the same direction.

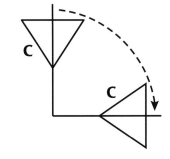

Triangle Y has been translated to the right.

Rotation—turning a figure around a particular point along a circular path.

Triangle C has been rotated 90°.

Reflection—making a mirror image of a figure by flipping it over a line. This line is called the *line of reflection.*

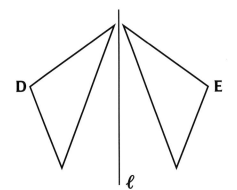

Triangle E is a reflection of triangle D over line ℓ.

5. a. Look at page 35. Choose two monkeys from Mia's picture and describe with these new terms (translation, rotation, and reflection) how you can move one monkey directly on top of the other. Repeat this problem with another pair of monkeys.

b. Refer to page 36. Describe your findings for problems **2** and **3** using rotations, translations, and reflections.

4. Yes, they are congruent. Explanations will vary. Students may trace one of the shapes and cut out the figure. They can then place it on top of the other shape. Students may also measure sides and angles to compare the triangles.

5. a. Answers will vary. Rotations, reflections, and translations are all used in these figures. Sample responses:

Monkey **A** to monkey **B:** rotate and translate.

Monkey **A** to monkey **C:** reflect, rotate, and translate.

Monkey **A** to monkey **D:** translate and rotate.

Monkey **A** to monkey **E:** reflect, rotate, and translate.

b. Shapes 2 and 3 are reflections of shape 1. Shapes 2 and 3 are rotations of one another. Shape 4 is a reflection of shape 2 (and of shape 3, as well, if the folds were made perfectly perpendicular). Shape 4 is a rotation of shape 1.

Materials congruent triangles, optional (two per class); overhead projector, optional (one per class)

Overview Students learn the concept of congruent figures. Then they learn the mathematical expressions for turns, flips, and slides and use these new terms in descriptions.

About the Mathematics A translation is a movement of a figure in a certain direction over a certain distance. This movement can be indicated by a vector. (The concept of a vector is made explicit in the grade 8/9 unit *Going the Distance*.) A rotation is the result of turning a figure around a particular point. This point is called the center of rotation. A reflection is the result of flipping a figure over a line. This line is called the line of reflection.

Planning When students finish problem **4,** you may wish to discuss the concept of congruent figures. Ask students to explain in their own words how they can tell whether two figures are congruent. Then discuss the text on this page. You may want to use two triangles and an overhead projector to show the movements and introduce their formal names. Students may work on problems **4** and **5** individually. Problem **5** may be used as an informal assessment. It may also be assigned as homework.

Comments about the Problems

5. Informal Assessment This problem assesses students' ability to describe transformations (translation, rotation, and reflection) using words and/or diagrams. This problem may be assigned as homework.

These designs were created by moving one triangle to new locations and changing colors.

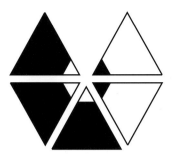 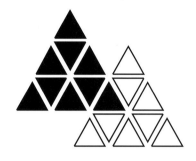 

Design A   **Design B**   **Design C**

6. a. For each design, pick a starting triangle and use the terms translation, rotation, and reflection to describe the movements that can be used to make the design.

b. Create your own design with triangles.

On page 36, you made two reflections of an irregular shape over two perpendicular lines. The lines that you folded the paper along are lines of reflection. You saw that these two reflections have the same result as rotating the original shape.

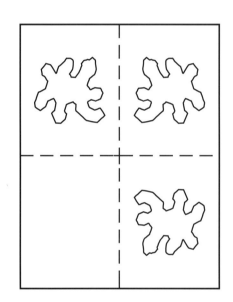

7. a. If the two lines of reflection are not perpendicular, is the resulting image still a rotation of the original shape? Use cutting and folding as you did on page 36 to discover what happens.

b. Describe the resulting image if the two lines of reflection are parallel.

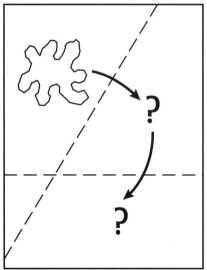

6. a. Answers will vary. Sample responses:

For design **A,** start with the upper left triangle and rotate it about its white vertex. Repeat three more times and change the coloring. Design **A** also could be made by reflecting one triangle over several lines as shown below.

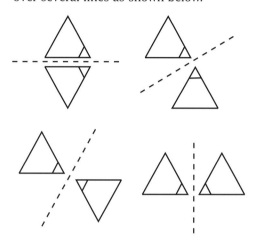

Designs **B** and **C** can be created by reflecting triangles after starting with one. These designs can also be made by translating larger groups of triangles.

b. Designs will vary. Sample design:

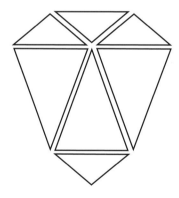

7. a. Yes, the resulting image is still a rotation.

b. Answers may vary. Students may note that in this case, the two reflections cannot be replaced by one rotation. They can be replaced by a translation as shown below:

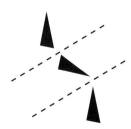

Materials drawing paper (one sheet per student); scissors (one pair per student)

Overview Students investigate designs that can be made by translating, rotating, and/or reflecting one figure. Then they create their own designs. Students investigate whether or not two reflections can always be replaced by a rotation.

About the Mathematics When a figure is reflected twice across intersecting lines, the result is the same as a rotation of the original figure. (And, actually, the angle of rotation is double the angle between the lines of reflection.) When a figure is reflected twice across parallel lines, the result is the same as a translation of the original figure (and the distance of the translation is twice the distance between the lines). Do not expect students to discover the specific measurement relationship. An even number of reflections will produce the original figure, and an odd number of reflections will produce a mirror image of the original figure.

Planning Discuss students' descriptions for problem **6a.** You may want to ask students to hang the creations they make for problem **6b** on a bulletin board. Problem **6** may be assigned as homework. Problem **7** is optional. Students may work on these problems individually.

Comments about the Problems

6. Homework This problem may be assigned as homework.

a. Some students may have used a reflection; others may have used a rotation for the same situation. This may lead to the statement that a reflection of an equilateral triangle can be replaced by a rotation and vice versa.

7. This problem is optional. If time is a concern, you may omit this problem or assign it as homework. Alternatively, you may want to divide the class into two groups, with one group working on part **a** and the other on part **b.** Then have the two groups exchange their results.

Reflection is often used in art. In these two pictures of cats, there is one line of reflection, but there is a difference between the pictures.

8. What do you think this difference is?

A B C D E F G H I J K L M N O P Q R S T U V W X Y Z

If a figure contains a line of reflection, it has *line symmetry*.

9. Which capital letters have line symmetry?

Some symmetrical shapes have more than one *line of symmetry*, as shown in the figure on the right.

10. Which capital letters have more than one line of symmetry?

11. If possible, design a triangle with the following characteristics:

    a. one line of symmetry

    b. two lines of symmetry

    c. three lines of symmetry

12. In your notebook, draw a rectangle, a rhombus, and a square and draw their lines of symmetry. How are their lines of symmetry different?

13. Create a design that has one or more lines of symmetry.

8. The second picture contains two separate images that are reflections of each other. The first picture is a single image that has a reflection within it.

9. A, B, C, D, E, H, I, M, O, T, U, V, W, X, and Y have line symmetry.

10. H, I, O, and X

11. a. An isosceles triangle that is not an equilateral triangle has exactly one line of symmetry as shown below.

b. No triangle has exactly two lines of symmetry.

c. An equilateral triangle has three lines of symmetry, as shown below.

12. As shown below, a rectangle has two lines of symmetry, a rhombus has two lines of symmetry, and a square has four lines of symmetry.

rectangle   rhombus

square

13. Designs will differ. Sample design:

Eight lines of symmetry:

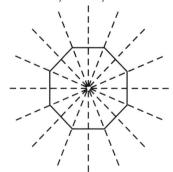

Materials hand-held mirror, optional (one per class)

Overview Students identify symmetry in figures and look for one or more lines of symmetry in these figures. They also create figures with one or more lines of symmetry.

About the Mathematics A figure that contains a line of reflection has line symmetry. This means that if you connect any two matching points, the line of symmetry would be the perpendicular bisector of the segment drawn.

Planning You may want to discuss problem **8** before students continue, to check that they understand the concept of line symmetry. Have students work individually on problems **8–10.** Then they can work in small groups on problems **11–13.** Problem **9** may be used as an informal assessment.

Comments about the Problems

8. The reflection may be difficult for students to see. Note that line symmetry also can be verified by folding the figure along the line of symmetry. Both parts of the figure should cover each other perfectly.

9. Informal Assessment This problem assesses students' ability to understand and identify line symmetry.

11. To solve this problem, students must be able to recognize line symmetry, create a symmetrical figure, explain why it is symmetrical, and test that symmetry.

12. Notice that a square has elements of both the rectangle and the rhombus. It has horizontal and vertical symmetry like a rectangle and diagonal symmetry like a rhombus.

13. All students should be able to design a figure with one line of symmetry. If students find this difficult, you may want to remind them that the folding line is the line of symmetry.

Extension The right side of the human body is a (near) reflection of the left side. You can demonstrate this by holding a mirror up to half of a person's face and having students observe by standing at an angle to the mirror.

Summary

Two figures are congruent if they are exactly the same size and shape.

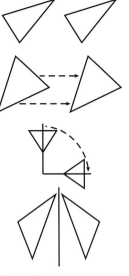

When you slide a figure so that each point moves the same distance in the same direction, it is called a translation.

A rotation means that a figure is turned around a point along a circular path.

A reflection is the process of making a mirror image by flipping a figure over a line. A figure that contains a line of reflection has line symmetry.

If a figure is translated, rotated, or reflected, the resulting figure is congruent to the original figure.

Summary Questions

14. Draw a figure and show how to translate, rotate, and reflect it.

15. Write a few sentences describing what it means for something to be symmetrical. You may want to include examples of symmetrical objects.

14. Drawings will vary. Sample drawing:

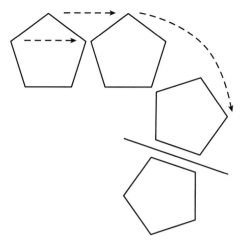

a translation, a rotation, and a reflection

15. Answers will vary. Sample student responses:

For anything to be symmetrical, there has to be at least one line of symmetry. A line of symmetry splits an object in half so that if you could fold the object along this line, each side would match up perfectly.

Put a mirror along the line of symmetry and see if the original part and the reflected part are the same.

Many everyday objects are symmetrical. For example, the fork and a spoon pictured below each have one line of symmetry.

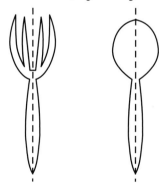

A cup with a handle has only one line of symmetry, a cup without a handle has many lines of symmetry as shown in the drawings:

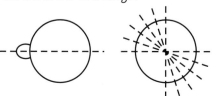

Overview Students read the Summary, which describes the terms *congruent, translation, rotation, reflection,* and *line symmetry.* Then they solve two problems using these concepts.

Planning Have students read the Summary and work on problems **14** and **15** individually. These problems may be used as informal assessment. After students complete Section F, you may assign appropriate activities from the Try This! section, located on pages 45–48 of the *Triangles and Beyond* Student Book, for homework.

Comments about the Problems

14. Informal Assessment This problem assesses students' ability to describe transformations (translation, rotation, and reflection) using words and/or diagrams.

15. Informal Assessment This problem assesses students' ability to understand and identify line symmetry.

Work Students Do

In this section, students rotate triangles to find out what polygons can be created and investigate the relationship between the triangle and the polygon that was created with it. In another activity, students rotate and then translate a copy of a triangle in such a way that they create a parallelogram when the two triangles are put together. Students investigate relationships between the triangle and the type of parallelogram formed.

Goals

Students will:

- use properties of triangles (sum of angles, side relationships, Hinge theorem);*
- identify congruent figures;
- describe geometric figures using words and/or diagrams;
- describe transformations (translation, rotation, and reflection) using words and/or diagrams.*

  * These goals are assessed in other sections of the unit.

Pacing

- approximately two 45-minute class sessions

Vocabulary

- vertex angle

About the Mathematics

This section pulls many of the activities and concepts of the previous sections together in order for students to further investigate geometric figures like polygons and parallelograms.

It is possible to make any regular polygon by rotating a triangle with the correct vertex angle, which must be a factor of 360. For example, since 60 goes into 360 six times, you can rotate a triangle with a vertex angle of 60° five times to create a hexagon.

The characteristics of parallelograms are emphasized when they are constructed from a triangle and its image. This image is the result of a rotation of the original triangle around one of the vertices, followed by a translation, as shown below:

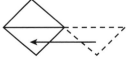

Materials

- Student Activity Sheet 8, page 132 of the Teacher Guide (one per student)
- drawing paper, pages 99 and 103 of the Teacher Guide (two sheets per student)
- scissors, pages 99 and 103 of the Teacher Guide (one pair per student)
- compasses or protractors, pages 99 and 101 of the Teacher Guide (one per student)
- rulers, pages 99, 101, and 103 of the Teacher Guide (one per student)
- colored pencils, page 103 of the Teacher Guide (one box per student)

Planning Instruction

You may want to begin this section with a class discussion about isosceles triangles, making sure students understand the concept of the vertex angle. Then ask students to give examples of regular polygons. Students may remember polygons from the unit *Packages and Polygons.*

Students may work on problems 1–4, 6, and 7 in small groups. Students may work individually on problems 5 and 8.

There are no optional problems in this section.

Homework

Problems 3 and 4 (page 100 of the Teacher Guide) can be assigned as homework. The Extensions (pages 101 and 103 of the Teacher Guide) may also be assigned as homework. After students complete Section G, you may assign appropriate activities from the Try This! section, located on pages 45–48 of the *Triangles and Beyond* Student Book. The Try This! activities reinforce the key mathematical concepts introduced in this section.

Planning Assessment

- Problems 5 and 8a can be used to informally assess students' ability to describe geometric figures using words and/or diagrams.
- Problem 8b can be used to informally assess students' ability to identify congruent figures.

Activity

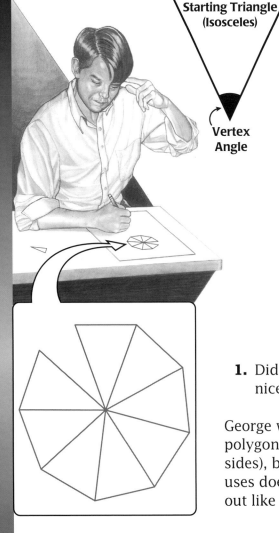

Starting Triangle (Isosceles)

Vertex Angle

i. Draw an isosceles triangle and color in the *vertex angle*, the angle that is formed by the two equal sides.

ii. Cut the triangle out and trace it on a sheet of paper.

iii. Rotate the triangle around the vertex, as shown below.

iv. Trace your triangle and rotate it again.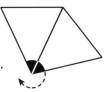

v. Continue this process until you have eight copies.

1. Did all eight copies fit together nicely, with no gaps or overlaps?

George wants to make an octagon (a polygon with eight angles and eight sides), but the isosceles triangle that he uses does not work. His drawing turns out like the one on the left.

2. a. Make a triangle that will form an octagon when you rotate and trace it eight times.

b. How can you decide in advance whether or not rotating and tracing a triangle will form an octagon?

c. Make a triangle that will form a hexagon when you rotate and trace it six times.

d. Can any regular polygon be created by rotating and tracing an isosceles triangle? If so, how?

1. Answers will vary. The eight copies will not fit together unless the vertex angle of the isosceles triangle is 45°.

2. a.

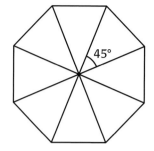

b. The vertex angle must be 360° ÷ 8 = 45°.

c.

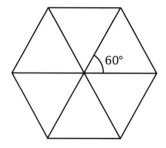

The vertex angle must be 360° ÷ 6 = 60°.

d. Yes. Explanations may vary. The following polygons can be made with isosceles triangles that have the indicated vertex angles:

Regular Polygon	Vertex Angle
Square	360° ÷ 4 = 90°
Pentagon	360° ÷ 5 = 72°
Hexagon	360° ÷ 6 = 60°
Nonagon	360° ÷ 9 = 40°
Decagon	360° ÷ 10 = 36°

Any regular polygon can be created from an isosceles triangle by determining the size of the vertex angle and then constructing the isosceles triangle accordingly.

Materials drawing paper (two sheets per student); scissors (one pair per student); compasses or protractors (one per student); rulers (one per student)

Overview Students cut an isosceles triangle out of a piece of paper and rotate the triangle. Then they find out what triangle, when rotated seven times, can make an octagon and what triangle can make a hexagon after five rotations. Students then investigate other regular polygons that can be created by rotating and tracing isosceles triangles.

About the Mathematics Students may remember the regular polygons from the unit *Packages and Polygons.* Any regular polygon can be divided into a number of congruent isosceles triangles. The measurement of the vertex angle of one such triangle can be found by dividing 360° by the number of triangles.

Planning For an idea of how to begin this section, see the Planning Instruction section in the section opener. Students may work on the activity and on problems **1** and **2** in small groups. Then have students share their findings with the whole class.

Comments about the Problems

2. a. Students should use trial and error if necessary. You may want students to explain to each other how they found their solutions. Some students may say that an octagon can be made from eight triangles; half the octagon has four triangles; and half of that has two triangles. The vertex angle of these two triangles together is 90°:

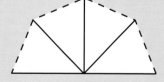

d. Students may answer with specific examples or may generalize with a high level of understanding. It is important that students see that this process is not limited to hexagons and octagons. If time permits, let students experiment with some other regular polygons.

3. a. On **Student Activity Sheet 8,** draw a triangle in each regular polygon that can be rotated to create the polygon.

 b. What is the measure of the vertex angle for each triangle that you drew?

Figure 1   **Figure 2**   **Figure 3**

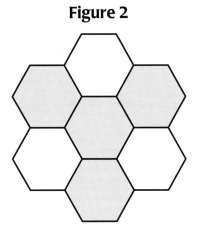   

4. a. Look at the figure on the right. Which two of the three movements (translation, rotation, or reflection) can be made with triangle A to result in triangle B?

 b. If you continue this pattern until you have a 10-gon, does it matter which of the two movements you use?

 c. Notice that triangle A is an isosceles triangle. Do the two movements you selected in part **a** produce the same result if the triangle is not isosceles? Why or why not? (You may investigate this by cutting out a triangle that is not isosceles, such as the one shown on the right, and making the two movements.)

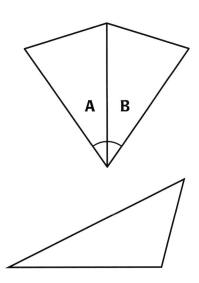

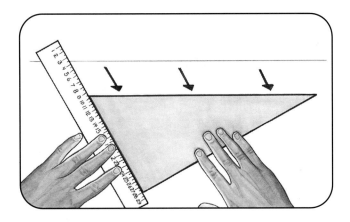

Recall that a translation is similar to sliding a figure along a straight line. When you made parallel lines with the triangle in Section E, you used a translation.

3. a.

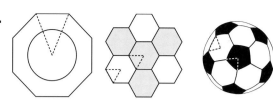

b. Figure 1: 360° ÷ 8 = 45°
Figure 2: 360° ÷ 6 = 60°
Figure 3: 360° ÷ 6 = 60° for triangles in hexagon
　　　　　360° ÷ 5 = 72° for triangles in pentagon

4. a. reflection or rotation

b. no

c. No. If the starting triangle is not isosceles, the reflection will give a different result. A rotation also will give a different result. You will not end up with a regular polygon with either movement, as shown below.

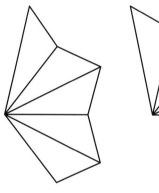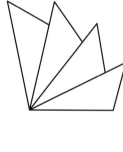

　　　　reflection　　　　　　　　rotation

Materials Student Activity Sheet 8 (one per student); compasses or protractors (one per student); rulers (one per student)

Overview Students draw triangles in polygons and find the measure of the vertex angles. They investigate whether translation, rotation, or reflection can be used to make a 10-gon. Then they investigate whether or not they would get the same result if they started with a triangle that is not isosceles.

About the Mathematics In general, any symmetric shape that is rotated about a point that lies on the line of symmetry can be replaced by a reflection. The reflection line becomes a line of symmetry of the original and its image:

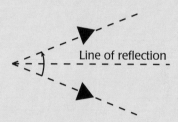

Line of reflection

Planning Have students work on problems **3** and **4** in small groups. These problems may also be assigned as homework.

Comments about the Problems

3–4. Homework These problems may be assigned as homework.

4. c. Some students may find it easier to explain by using a drawing of what the movements would produce.

If the starting triangle is not isosceles, a reflection over a side will change the orientation; a rotation will provide the same orientation but will not form a regular polygon.

Extension You may want students to draw pictures of the reflections and rotations of the triangle described in problem **4c.**

Activity

For this activity, you will combine a rotation and a translation.

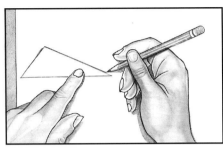

i. Make a cutout of any triangle and trace it on a piece of paper.

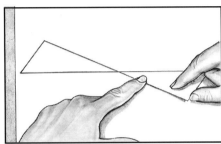

ii. Rotate the triangle 180° around one vertex until it is upside down, as in the picture on the left.

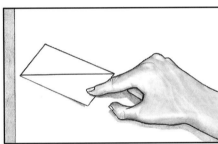

iii. Translate the triangle so that it lines up alongside or below the starting triangle and trace this final position.

5. a. What kind of quadrilateral have you made?

   b. Color the angles of your quadrilateral so that angles with the same measures are the same color.

   c. Repeat steps i through iii with a different triangle.

   d. What conclusions can you draw?

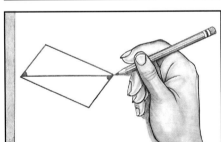

6. If you want to use a rotation and a translation to make a rhombus, what kind of triangle should you start with?

7. What kind of triangle should you rotate and translate to make a rectangle?

5. a. a parallelogram

b–c. Figures will vary, depending on the triangle chosen. Sample responses:

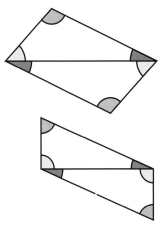

d. Conclusions will vary. Sample conclusions:

This process always creates a parallelogram. Opposite angles in a parallelogram are equal.

6. an isosceles triangle, as shown below:

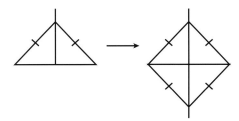

7. a right triangle, as shown below:

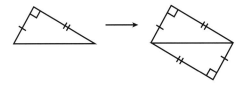

Notice that if you start with the triangle below,

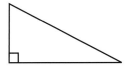

and rotate it, the translation should be at an angle rather than horizontal, as shown below.

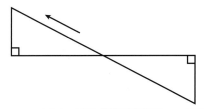

Materials drawing paper (two sheets per student); scissors (one pair per student); colored pencils (one box per student); rulers (one per student)

Overview Students rotate and then translate a triangle to make a quadrilateral. They find out what type of triangle can be used to make a rhombus or a rectangle.

About the Mathematics Two angles in a quadrilateral are either consecutive or opposite.

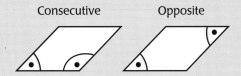

In a parallelogram, these angle pairs are special. From problem **5b,** students should understand that opposite angles of a parallelogram are equal. The transformation construction also shows that in a parallelogram, the sum of consecutive angles is 180°.

Planning Students may work individually on problem **5.** This problem may be used as an informal assessment. Students may work in small groups on problems **6** and **7.**

Comments about the Problems

5. Informal Assessment This problem assesses students' ability to describe geometric figures using words and/or diagrams.

b. Ask students, *How can you be sure that the angles you chose are equal?* [because they were made from the same triangle]

5–7. These problems are important for the development of students' understanding of parallelograms and their characteristics. *Note:* The results of students' investigations may help them see that a diagonal of a parallelogram does not necessarily divide adjacent angles equally.

Extension You may want to ask students the following questions: *Can you rotate a triangle around any vertex to make a parallelogram?* [yes] *Will the result always be the same after you have translated the triangle?* [no] Encourage students to find the answers by making drawings.

Summary

A rotation and a reflection sometimes produce the same result. For example, when the triangle is isosceles, a rotation around the vertex formed by the equal sides is the same as a reflection over one of the equal sides.

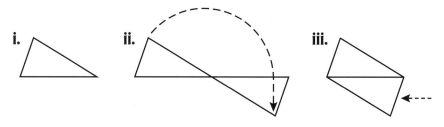

rotation   **reflection**

Regular polygons can be made by rotating isosceles triangles. The measure of the vertex angle of an isosceles triangle determines whether or not that triangle will form a regular polygon. The angle measure must be a factor of 360°.

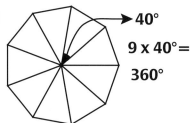

40°

9 x 40°=

360°

By combining a rotation and a translation, any triangle can be used to draw a parallelogram.

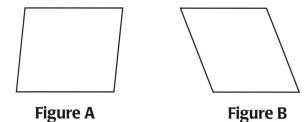

i.   **ii.**   **iii.**

Summary Questions

8. Study the two shapes shown below. Describe several ways to determine the following:

    a. if each shape is a parallelogram

    b. if the two shapes are congruent

Figure A   **Figure B**

8. a. Answers will vary. Sample responses:

I would see if the opposite sides are parallel.

I would divide the quadrilateral into two triangles. Then I would see whether or not one triangle can be moved by a rotation and a translation so that it fits on the other triangle.

I would measure the angles and compare them. If the opposite angles are equal, then it is a parallelogram.

I would measure the sides. If the opposite sides are equal, then it is a parallelogram.

b. Answers will vary. Sample responses:

I would measure the sides and the angles. If they match, then the shapes are congruent.

I would cut out one shape and see whether or not it fits on the other.

Overview Students read and discuss the Summary. Then they investigate whether or not two shapes are parallelograms and whether they are congruent.

Planning Students may work individually on problem **8.** This problem may be used as informal assessment. After students complete Section G, you may assign appropriate activities from the Try This! section, located on pages 45–48 of the *Triangles and Beyond* Student Book, for homework.

Comments about the Problems

8. You may want students to write their descriptions in their notebooks before answering.

a. Informal Assessment This problem assesses students' ability to describe geometric figures using words and/or diagrams.

b. Informal Assessment This problem assesses students' ability to identify congruent figures.

Assessment Overview

Students work on five assessment activities that you can use to collect additional information about what each student knows about properties of triangles, parallelograms, and symmetry.

Goals	Assessment Opportunities
• recognize triangles in the world	Do You See the Mathematics?
• use properties of triangles (such as the sum of angles, side relationships, and the Hinge Theorem)	The Dragline
• construct a triangle with given side lengths	The Rolling Triangle
• use construction when appropriate	The Dragline
• describe geometric figures using words and/or diagrams	Do You See the Mathematics? Rhombus Copies The Rolling Triangle
• describe transformations (translation, rotation, and reflection) using words and/or diagrams	Do You See the Mathematics? Copies The Rolling Triangle
• appreciate that geometry is a means of describing the world	Do You See the Mathematics? The Dragline
• create mathematical representations from visualization	Copies The Rolling Triangle
• recognize the need for mathematical rigor in justifying answers	Rhombus

Pacing

When combined, the five assessment activities will take approximately two 45-minute class sessions. See the Planning Assessment section for further suggestions as to how you might use the assessment activities.

About the Mathematics

The five assessment activities evaluate the major goals of the *Triangles and Beyond* unit. Refer to the Goals and Assessment Opportunities section on the previous page for information regarding the specific goals assessed in each assessment activity. Students may use different strategies to solve each problem. Their choice of strategies may indicate their level of comprehension of the problem. Consider how well students' strategies address the problem, as well as how successful students are at applying their strategies in the problem-solving process.

Materials

- Assessments, pages 133–139 of the Teacher Guide (one of each per student)
- index cards, pages 109 and 111 of the Teacher Guide (three per student)
- scissors, pages 109, 111, 113, and 117 of the Teacher Guide (one pair per student)
- centimeter rulers, pages 109 and 111 of the Teacher Guide (one per student)
- compasses, pages 109 and 121 of the Teacher Guide (one per student)
- compass cards or protractors, page 111 of the Teacher Guide (one per student)
- transparent paper, pages 113 and 115 of the Teacher Guide, optional (one sheet per student)
- colored pencils, pages 113 and 115 of the Teacher Guide (one box per student)
- glue or tape, page 117 of the Teacher Guide, optional (one dispenser per student)
- 15-cm long paper strips, page 121 of the Teacher Guide, optional (one strip per group of students)

Planning Assessment

You may want students to work on these assessment problems individually if you want to evaluate each student's understanding and abilities. Make sure that you allow enough time for students to complete the assessment activities. Students are free to solve each problem in their own ways. They may choose to use any of the models introduced and developed in this unit to solve problems that do not call for a specific model.

Scoring

Answers are provided for all assessment problems. The method of scoring the problems depends on the types of questions in each assessment. Most questions require students to explain their reasoning or justify their answers. For these questions, the reasoning used by the students in solving the problems as well as the correctness of the answers should be considered as part of your grading scheme. A holistic scoring scheme can be used to evaluate an entire task. For example, after reviewing a student's work, you may assign a key word such as *emerging, developing, accomplishing,* or *exceeding* to describe his or her mathematical problem solving, reasoning, and communication.

On other tasks, it may be more appropriate to assign point values for each response. Students' progress toward the goals of the unit should also be considered. Descriptive statements that include details of a student's solution to an assessment activity can be recorded. These statements would provide insight into a student's progress toward a specific goal of the unit. Descriptive statements can be more informative than recording only a score and can be used to document students' growth in mathematics over time.

Use additional paper as needed.

Jade wants to decorate the walls of her bedroom. She wants the border to be in a pattern of triangles. To see what her pattern will look like, do the following:

Construct triangle *ABC* with *AB* = 6 centimeters, *BC* = 4 centimeters, and *AC* = 5 centimeters on an index card or thick paper. Cut out the triangle and place side *AB* on a line as shown below.

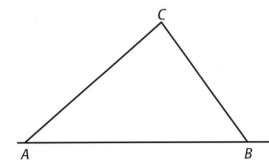

1. To show the rolling process, start with the triangle in the position shown above. Trace the triangle, roll it once, and trace it again. Roll the triangle again and trace it. After each roll, one side of the triangle should be on the line.

2. You should have three triangles on your line. Do you need to continue drawing triangles to see the next position? Explain your answer.

3. Find the distance on the line covered by triangle *ABC* after three rolls.

4. Trace the movement of the three endpoints (*A*, *B*, and *C*) during the rolling process of the first roll. Describe what you see. What movement do the three endpoints make?

1.

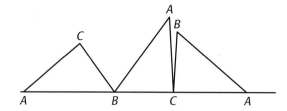

2. No. The fourth triangle will be in the same position as the first because a triangle has only three sides.

3. 6 cm + 4 cm + 5 cm = 15 cm

4.

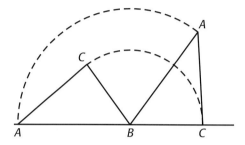

Answers may vary. Sample responses:

The paths are curves.

The movement of the points *A* and *C* is a rotation around point *B*. Point *B* doesn't move during the first roll.

Materials The Rolling Triangle assessment, pages 133–134 of the Teacher Guide (one of each per student); index cards (three per student); scissors (one pair per student); centimeter rulers (one per student); compasses (one per student)

Overview Students construct a triangle. Then they create a pattern of triangles by rolling the triangle along a line and tracing the result.

About the Mathematics These assessment activities evaluate students' ability to construct a triangle with given side lengths and to describe geometric figures using words and/or diagrams. They also assess students' ability to describe rotations and to create mathematical representations from visualization.

Planning You may want students to work on these assessment problems individually.

Comments about the Problems

3. Some students may want to measure the distance in their drawing. Encourage them to solve this problem without measuring.

4. The curves are actually parts of circles, because they are traced by outlining all of the points at a fixed distance from the rotation point. You may want to have students use compasses to make the curves.

Use additional paper as needed.

5. If you had rolled triangle *ABC* a third time, how many degrees would the triangle have been rotated in total?

6. Roll an isosceles triangle and an equilateral triangle along a line. Continue rolling and tracing the triangle until your design is long enough to show a pattern. What differences do you notice between these triangle patterns and the one Jade used?

5. 360°. Three rolls will result in one complete turn of the triangle.

6. Answers will vary. Sample responses:

The pattern created by an equilateral triangle repeats after every rotation, while the patterns of the other two repeat after three rotations.

The patterns with the isosceles and equilateral triangles have more symmetry than the pattern made with the scalene triangle, as shown below.

Materials The Rolling Triangle assessment, pages 133–134 of the Teacher Guide (one of each per student); index cards (three per student); scissors (one pair per student); centimeter rulers (one per student); compass cards or protractors (one per student)

Overview Students repeat the rolling process two more times, first with an isosceles triangle and then with an equilateral triangle. They compare the results and describe the differences.

Comments about the Problems

6. Some students may describe the variation in heights of the patterns. Other students may explain that the patterns do not have to be made by rotating the triangles. For example, the pattern made by rotating the equilateral triangle can also be made by translating the triangle.

Extension You may want to have students take another look at the first pattern they made. Students should measure the angles of the triangles they used. Ask students, *How many degrees did the triangle turn during the first roll?* [Angle $A = 40°$, angle $B = 54°$, and angle $C = 86°$. The rotation of the first roll is $180° -$ angle B, or $180° - 54° = 126°$.]

Use additional paper as needed.

All the triangles in the picture below are copies of triangle 1. They were made using rotation, translation, reflection, and a combination of these moves.

1. Describe how you could move triangle 1 to make copies 2, 3, 4, and 5.

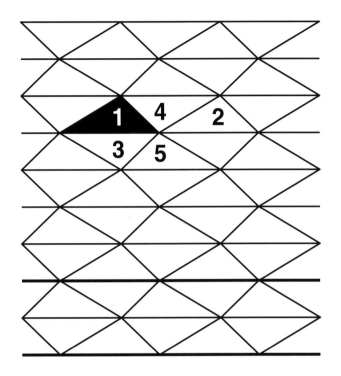

2. Two of the angles in triangle 1 are 42° and 108°.

    a. What is the measure of the third angle in triangle 1?

    b. What are the measures of the angles in triangles 2 and 4?

    c. Triangles 1, 2, and 4 meet at a point where three angles touch. What geometric property of triangles is shown where the three angles touch?

3. Label the 10 triangles between the thick lines with the letters T, R, or M based on the following rules:

    T, if you can get to the exact position of the triangle with a translation from triangle 1.

    R, if you need a rotation of triangle 1 along with a translation to get to the exact position of the triangle.

    M, if you need a reflection of triangle 1 combined with a translation.

1. Answers will vary. Sample responses:

Triangle 2 is a translation of triangle 1.

Triangle 3 is a reflection of a triangle 1.

Triangle 4 is a rotation and translation of triangle 1.

Triangle 5 is a reflection, a rotation, and a translation of triangle 1.

2. a. 30°. Sample strategy:

42° + 108° = 150°.

180° − 150° = 30°.

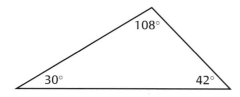

b. The angles are the same as the angles in triangle 1 (42°, 108°, and 30°).

c. The sum of the angles in a triangle is 180°. Together, the three angles form a straight line, which measures 180°.

3.

Materials Copies assessment, pages 135–136 of the Teacher Guide (one of each per student); transparent paper, optional (one sheet per student); scissors, optional (one pair per student); colored pencils, optional (one box per student)

Overview Students investigate how a triangle can be moved to create a design. They describe the movement using the terms *rotation, reflection,* and *translation.* Students also determine measurements of angles.

About the Mathematics These assessment activities evaluate students' ability to describe geometric figures using words and/or diagrams, to describe transformations, and to create mathematical representations from visualization.

Planning You may want students to work on these assessment problems individually.

Comments about the Problems

1. If students have difficulty, you might have them trace the triangle, cut it out, and move it to make the copies.

2. a. Students should not measure the angle with a protractor, but should find the measure through reasoning.

c. You might have them color the angles, so that equal angles in the picture each get the same color. Then, students should see that each of the three angles that meet have a different color. Thus, these three angles must represent the three angles of the first triangle.

3. The positions of the triangles that are labeled with an R in the solutions column could also be arrived at with just one rotation:

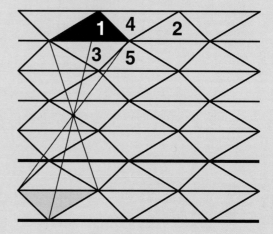

Use additional paper as needed.

The sides of the triangles pictured on the right belong to families of parallel lines. Two members of one family have been indicated in the drawing.

4. How many different families are there? Identify three members of each family using the same family letter.

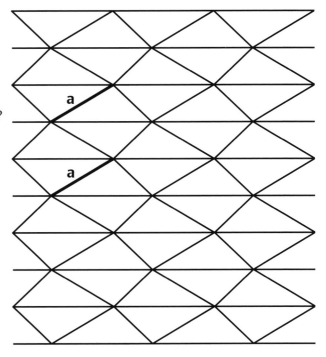

The triangles in this figure are arranged so you can see many parallelograms.

5. Use colored pencils to outline the various parallelograms in the figure.

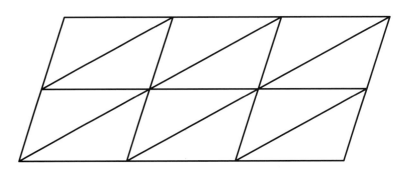

4. There are five families of parallel lines.

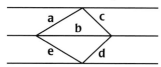

One way to label three members of each family is as follows:

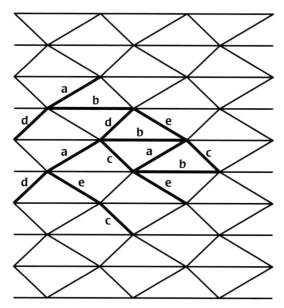

5. There are nine parallelograms in this figure, including the whole figure and eight other parallelograms:

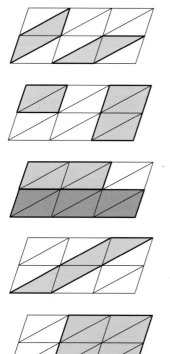

Materials Copies assessment, pages 135–136 of the Teacher Guide (one of each per student); transparent paper, optional (one sheet per student); colored pencils (one box per student)

Overview Students identify families of parallel lines. They also look for different-shaped parallelograms in a drawing of triangles arranged into parallelograms.

About the Mathematics These assessment activities evaluate students' ability to describe geometric figures using words and/or diagrams, describe transformations, and create mathematical representations from visualization.

Planning You may want students to work on these assessment problems individually.

Comments about the Problems

4. Students may use letters to mark parallel lines, as shown in the Solutions and Samples column, or they may use squiggly lines, hash marks, or other ways of identifying parallel lines.

5. The copies that are drawn in this picture are made by rotation and translation. Students can find a parallelogram by looking at one triangle and a copy that shares one of the sides (see the first three solutions). Another way is to look at quadrilaterals that are formed by two families of parallel lines.

RHOMBUS

Use additional paper as needed.

One way to determine whether or not quadrilateral *ABCD* is a rhombus would be to cut it out and fold it.

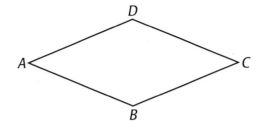

1. Before you do this, write down what you need to know about this quadrilateral in order to conclude that it is a rhombus.

2. Describe how you can use paper-folding to reach this conclusion.

Quadrilateral *EFGH* is a rhombus. It looks like two equilateral triangles have been put together.

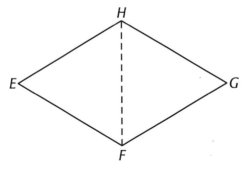

3. How can you use paper-folding to show that triangle *EFH* is equilateral?

1. For a quadrilateral to be a rhombus, all of its sides must be equal.

2. If you fold the figure over line *AC* and the sides match up, then *AD* = *BA* and *DC* = *BC*.

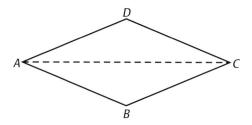

If you fold the figure over *DB* and the sides match up, then *AD* = *DC* and *AB* = *BC*.

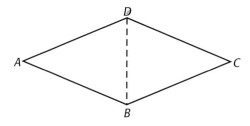

If you do both and the sides match both times, then *AD* = *BA* = *BC* = *DC*.

3. Answers will vary. Sample student response:

Since the shape is a rhombus, I know that EF and EH have equal lengths. To show that EF and EH are equal to FH, I can fold along either dotted line to match EF or EH with line FH.

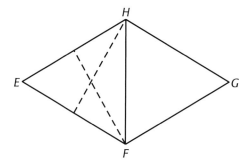

Materials Rhombus assessment, page 137 (one per student); scissors (one pair per student); glue or tape, optional (one dispenser per student)

Overview Students use paper-folding to determine whether a quadrilateral is a rhombus and whether a triangle within a rhombus is equilateral.

About the Mathematics These assessment activities evaluate students' ability to describe geometric figures using words and/or diagrams and to recognize the need for mathematical rigor in justifying answers.

Planning You may want students to work on these assessment problems individually.

Comments about the Problems

2–3. Allow students to actually cut out the shapes and make folding lines. They can paste or tape the shapes in their notebooks or on a piece of paper to show how they made their folding lines.

3. Students have already used the strategy of folding a shape twice to show line symmetry in problem **2.** They can use this strategy again to solve this problem. Students might reason as follows: if triangle *EFH* is an equilateral triangle, then it has three lines of symmetry. If you fold along these lines, each time you fold the two halves should match.

DO YOU SEE THE MATHEMATICS?

Use additional paper as needed.

1. Describe the familiar geometrical shapes and any transformations—such as reflections, rotations, translations, or combinations of the three—that might have been used to create the logos on this page.

a.

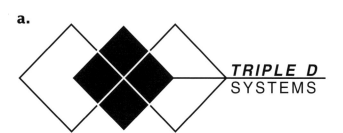

b.

Star Lighting Corp.

c.

d.

e.

f.

Central Outfitters

1. Answers will vary. Sample responses:

a. The logo is made from three congruent squares. The two outer squares divide the inner square into four congruent squares. The outer squares are reflections (or rotations) of one another. The inner squares are reflections (or rotations) and translations of one another.

b. There are eight congruent isosceles triangles; the inside space creates a star shape. To make this logo, you can start with one triangle, translate it, rotate it, translate, rotate, and so on, until you have made seven copies. You can also create the logo using all reflections. See the different reflection lines shown below:

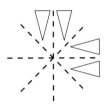

c. There are two large scalene triangles and one equilateral triangle (inside the A). One large triangle makes the other by a rotation.

d. There are one octagon and eight isosceles triangles. The octagon was formed by rotating or reflecting an isosceles triangle seven times.

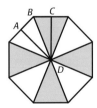

e. This logo is a parallelogram that includes two large congruent triangles (when it is cut horizontally) or two congruent parallelograms (when it is cut vertically). The large parallelogram was formed by rotating a large triangle (its base is the horizontal line) and translating it.

f. The arrow shapes are three rotations or three reflections of a single arrow. The black/white color pattern can be achieved by rotations or reflections of a shape formed by half of 2 arrows.

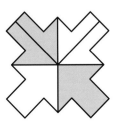

Materials Do You See the Mathematics? assessment, page 138 (one per student)

Overview Students describe which geometric shapes they recognize in the logos. Then they describe how transformations might have been used to create the logos.

Planning You may want students to work on these assessment problems individually. If time is a concern, you might have students work on only a few of these logos in class. The rest of the logos can be assigned as homework.

Bringing Math Home Students may explain to their families how logos can be made by starting with one shape and then using one or more transformations.

THE DRAGLINE

Use additional paper as needed.

A dragline is a machine that works like a crane and is used to dig in ponds or lakes. The most important part of the dragline is the long arm with a hinge at the top like an elbow.

Shown below are two schematic drawings of a dragline.

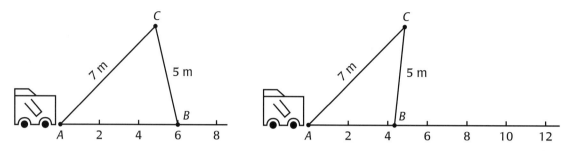

The dragline above has two arm sections, *AC* and *CB.* Point *A* is fixed to the machine, and *C* can move up or down. As *C* moves up or down, *B* moves either toward or away from the machine.

In the dragline above, arm section *AC* is 7 meters, and arm section *BC* is 5 meters.

1. Estimate the length of *AB* in the drawings above.

2. Use the scale above to draw a picture of when the arm sections of the machine reach out to 8 meters.

3. If you continue moving *B* to the right, the angle at *A* and the angle at *B* will become smaller. Will the angle at *A* always be smaller than the angle at *B*? Explain your answer.

4. **a.** Can the machine ever dig a hole 15 meters from *A*? Explain.

   b. Can the machine ever dig a hole 1 meter from *A*? Explain.

   c. Show in the drawing the maximum and minimum distances the arms can reach.

1. *AB* measures about 6 meters in the picture on the left and about 4.3 meters in the picture on the right.

2.

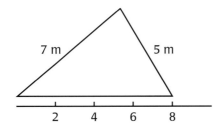

3. Yes, because *BC* is always smaller than *AC* and the smaller angle is opposite the smaller side.

4. a. No, because the two arms together are not longer than 12 meters.

 b. No, because if *AB* has to be one meter, then *AB* and *BC* together would not be longer than *AC*.

 c. The minimum distance is two meters. The maximum distance is 12 meters.

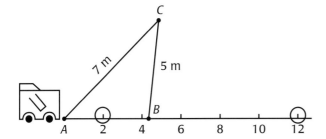

Materials The Dragline assessment, page 139 (one per student); 15-cm long paper strips, optional (one per group of students); compasses (one per student)

Overview Students investigate a dragline machine. They solve problems regarding the reaching distance of the machine.

Planning You may want students to work on these assessment problems in small groups.

Comments about the Problems

 1–4. These problems assess students' ability to recognize triangles in the world and to apply their knowledge about the properties of angles and sides of triangles. Students can use paper strips to simulate the machine.

 4. b. Students may use a variety of strategies to solve this problem. Some students may notice that to dig a hole one meter from *A, C* would have to go so far up that *B* could not reach the ground. Other students may be able to make connections with properties of triangles in their explanations. They may note that the total measurement of any two sides of a triangle is always more than the measurement of the third side.

Writing Opportunity You may want to ask students to write a report about their investigations of the dragline machine in their journals.

Triangles and Beyond
Glossary

The Glossary defines all vocabulary words listed on the Section Opener pages. It includes the mathematical terms that may be new to students, as well as words having to do with the contexts introduced in the unit. (*Note:* The Student Book has no glossary. This is in order to allow students to construct their own definitions, based on their personal experiences with the unit activities.)

The definitions below are specific to the use of the terms in this unit. The page numbers given are from this Teacher Guide.

congruent figures (p. 88) figures that are copies of each other; figures with the same size and shape; figures that can fit on top of each other exactly

equilateral triangle (p. 22) a triangle with sides of equal length; a triangle with three equal angles

family of parallel lines (p. 74) a set of lines that are all parallel to each other

isosceles triangle (p. 22) a triangle with at least two sides of equal length; a triangle with at least two equal angles

line of reflection (p. 88) a line over which a figure is reflected, so that a copy of the figure appears on the other side of the line; the line that lies directly between two mirror images

line of symmetry (p. 92) a line such that if a figure is folded on it, then one half of the figure matches the other half

line symmetry (p. 92) the inclusion of a line of reflection

parallel (p. 68) extending in the same direction, staying the same distance apart and never intersecting

parallelogram (p. 80) a quadrilateral formed by the intersection of two pairs of parallel lines

quadrilateral (p. 76) a four-sided figure

reflection (p. 88) a mirror image; a copy of a figure made by flipping it over a line

rhombus (p. 78) a parallelogram with four equal sides

rotation (p. 88) a copy of a figure made by turning the figure around a particular point along a circular path

scalene triangle (p. 22) a triangle with three sides of different lengths

translation (p. 88) a copy of a figure made by sliding the figure so that every point on the figure moves the same distance in the same direction

vertex angle (p. 98) the angle that is formed by the two congruent sides of an isosceles triangle

Blackline
Masters

Dear Family,

Your child will soon begin the *Mathematics in Context* unit *Triangles and Beyond*. Below is a letter to your child that opens the unit, describing the unit and its goals.

At the beginning of this unit, students are asked to find examples of triangles. You may want to help your child find pictures that include triangles in magazines and newspapers. After your child has worked through the unit, ask him or her to show you which triangles are equilateral, which are isosceles, and which are scalene.

You may also want to take a tour of your neighborhood and have your child point out some geometric shapes and properties. For example, your child should be able to show you a pair of parallel lines, such as two sidewalks running along opposite sides of the street.

We hope you and your child enjoy studying geometry together!

Dear Student,

Welcome to *Triangles and Beyond*.

Throughout this unit, you will study many kinds of triangles and parallelograms and their special geometric properties.

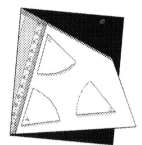

What happens if you try to make a triangle with three straws, one 12 centimeters long, one 4 centimeters long, and one 6 centimeters long? Do you think there is any connection between the location of the largest angle and the location of the longest side in a triangle? As you work with triangles, you will learn some requirements for the measures of their sides and angles.

You will also investigate the properties of parallel lines and learn the differences between parallelograms, rectangles, rhombuses, and squares.

As you work through this unit, look around you to see how the geometric shapes and properties you are studying appear in everyday objects.

Sincerely,

The Mathematics in Context Development Team

Sincerely,

The Mathematics in Context Development Team

3. Use one color to show three triangles that are part of the bridge's frame. Use another color to show three triangles that can only be seen in the drawing. You can check your answer by making a model of the bridge using toothpicks and clay or gumdrops.

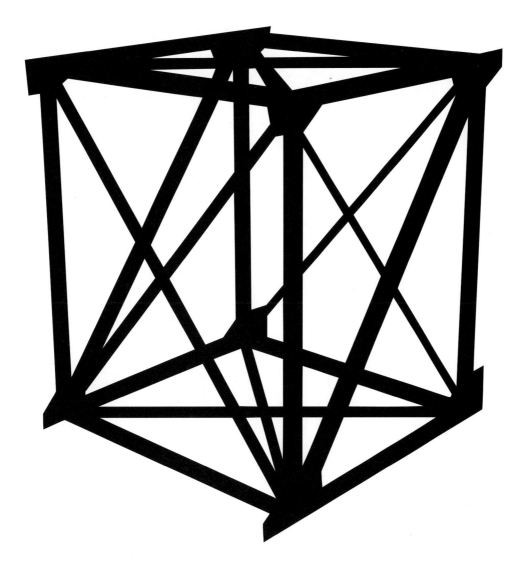

Use with *Triangles and Beyond,* page 5.

10. a. Draw the missing parts to restore the triangles shown below to their original shapes.

   b. Which of the triangles can be restored in more than one way? Why do you think only some of the triangles can be restored in more than one way?

A.
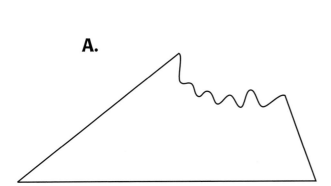

B.
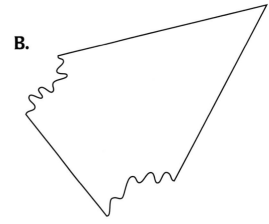

C.
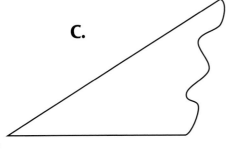

D.
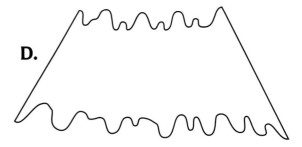

Use with *Triangles and Beyond,* pages 10 and 11.

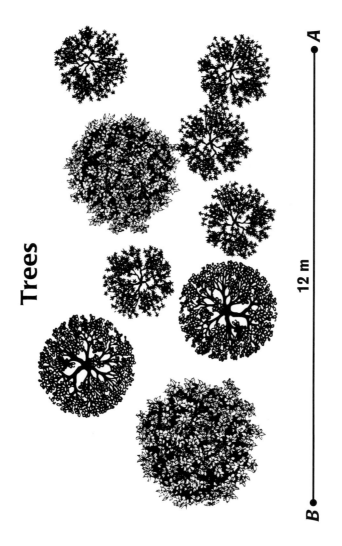

Trees

12 m

A

B

Name_____

Photo	Longest Side	Middle Side	Shortest Side
1			
2			
3			

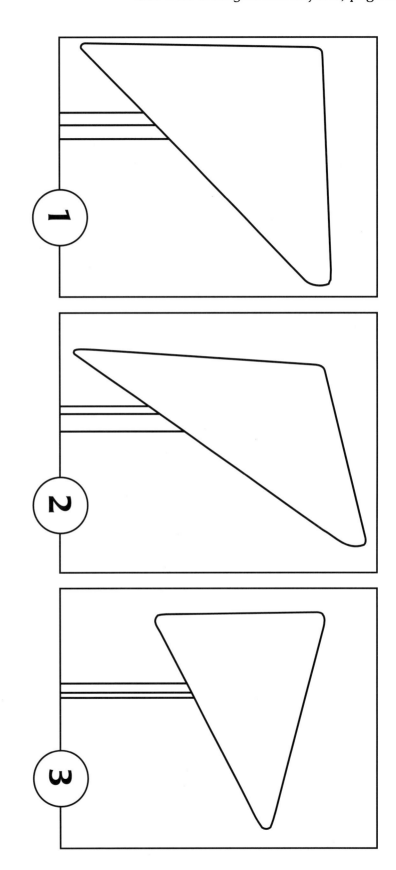

Use with *Triangles and Beyond,* page 23.

3. a. Measure the angles and sides of triangle *ABC* and label the drawing.

   b. Mark the triangle's largest angle.

   c. Mark the triangle's longest side.

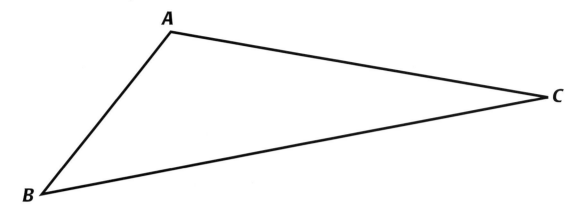

4. a. With one color, mark the smallest angles and the shortest sides of triangles I, II, and III.

   b. Next, with a different color, mark the largest angles and the longest sides of triangles I, II, and III.

   c. Do you see another possible geometric property of triangles from the answers to problems **3** and **4?** If so, what is the property?

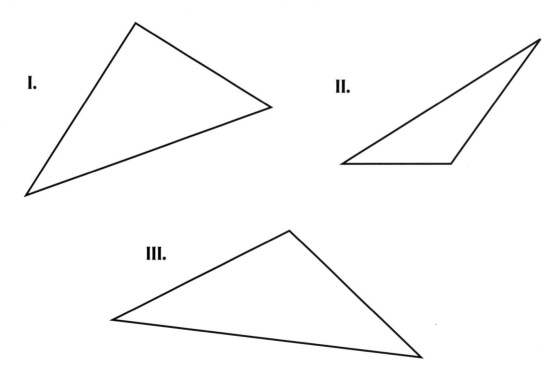

8. This truck is moving down a ramp into a tunnel. Locate the point at which the truck will be completely hidden from view by the wall. Show how you located this point.

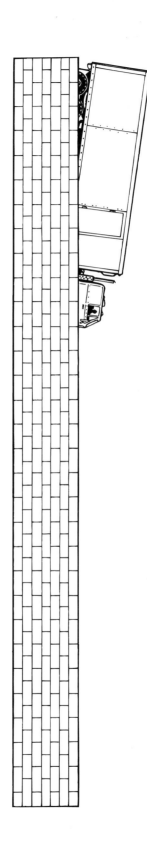

Student Activity Sheet 8

Name ______________________________

Use with *Triangles and Beyond*, page 42.

3. a. Draw a triangle in each regular polygon shown below that can be rotated to create the polygon.

b. What is the measure of the vertex angle for each triangle that you drew?

Figure 1

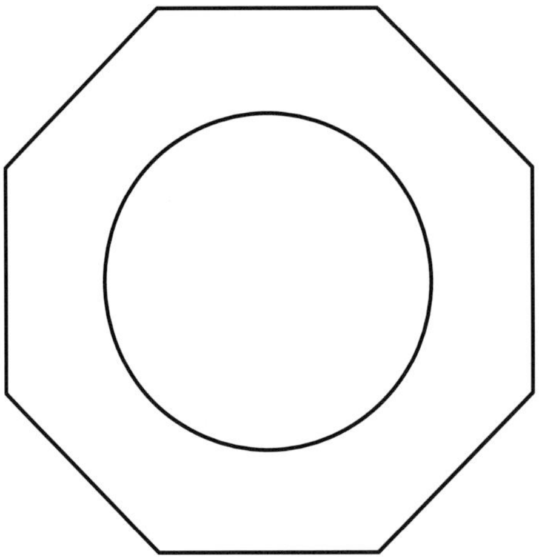

Figure 2

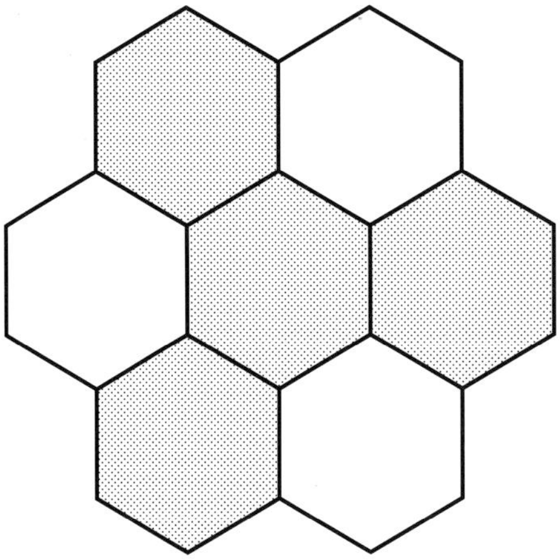

Figure 3

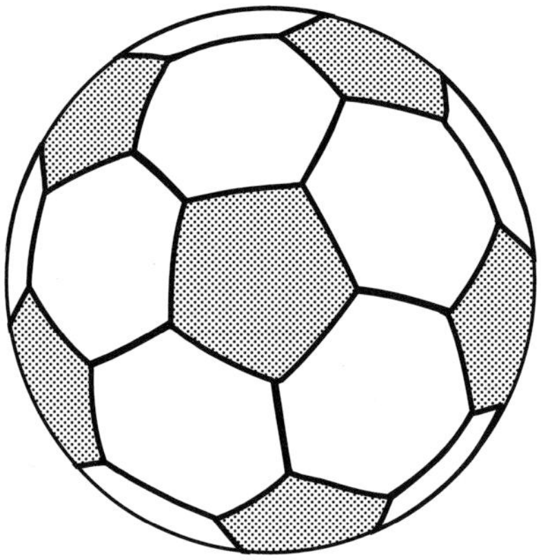

Use additional paper as needed.

Jade wants to decorate the walls of her bedroom. She wants the border to be in a pattern of triangles. To see what her pattern will look like, do the following:

Construct triangle *ABC* with *AB* = 6 centimeters, *BC* = 4 centimeters, and *AC* = 5 centimeters on an index card or thick paper. Cut out the triangle and place side *AB* on a line as shown below.

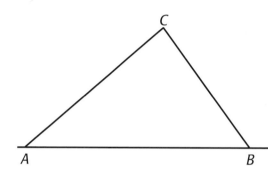

1. To show the rolling process, start with the triangle in the position shown above. Trace the triangle, roll it once, and trace it again. Roll the triangle again and trace it. After each roll, one side of the triangle should be on the line.

2. You should have three triangles on your line. Do you need to continue drawing triangles to see the next position? Explain your answer.

3. Find the distance on the line covered by triangle *ABC* after three rolls.

4. Trace the movement of the three endpoints (*A*, *B*, and *C*) during the rolling process of the first roll. Describe what you see. What movement do the three endpoints make?

Use additional paper as needed.

5. If you had rolled triangle *ABC* a third time, how many degrees would the triangle have been rotated in total?

6. Roll an isosceles triangle and an equilateral triangle along a line. Continue rolling and tracing the triangle until your design is long enough to show a pattern. What differences do you notice between these triangle patterns and the one Jade used?

Use additional paper as needed.

All the triangles in the picture below are copies of triangle 1. They were made using rotation, translation, reflection, and a combination of these moves.

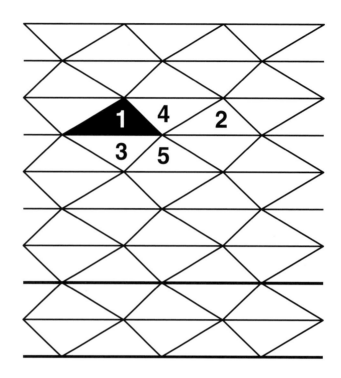

1. Describe how you could move triangle 1 to make copies 2, 3, 4, and 5.

2. Two of the angles in triangle 1 are 42° and 108°.

   a. What is the measure of the third angle in triangle 1?

   b. What are the measures of the angles in triangles 2 and 4?

   c. Triangles 1, 2, and 4 meet at a point where three angles touch. What geometric property of triangles is shown where the three angles touch?

3. Label the 10 triangles between the thick lines with the letters T, R, or M based on the following rules:

T, if you can get to the exact position of the triangle with a translation from triangle 1.

R, if you need a rotation of triangle 1 along with a translation to get to the exact position of the triangle.

M, if you need a reflection of triangle 1 combined with a translation.

COPIES

Use additional paper as needed.

The sides of the triangles pictured on the right belong to families of parallel lines. Two members of one family have been indicated in the drawing.

4. How many different families are there? Identify three members of each family using the same family letter.

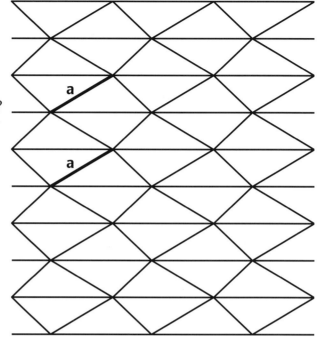

The triangles in this figure are arranged so you can see many parallelograms.

5. Use colored pencils to outline the various parallelograms in the figure.

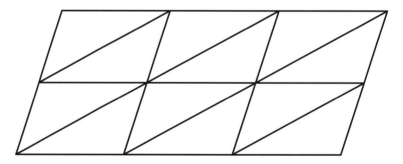

Use additional paper as needed.

One way to determine whether or not quadrilateral *ABCD* is a rhombus would be to cut it out and fold it.

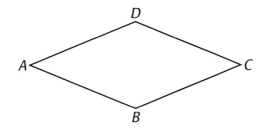

1. Before you do this, write down what you need to know about this quadrilateral in order to conclude that it is a rhombus.

2. Describe how you can use paper-folding to reach this conclusion.

Quadrilateral *EFGH* is a rhombus. It looks like two equilateral triangles have been put together.

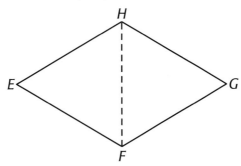

3. How can you use paper-folding to show that triangle *EFH* is equilateral?

DO YOU SEE THE MATHEMATICS?

Use additional paper as needed.

1. Describe the familiar geometrical shapes and any transformations—such as reflections, rotations, translations, or combinations of the three—that might have been used to create the logos on this page.

a.

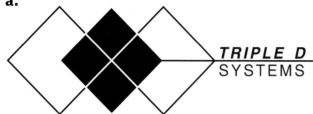

TRIPLE D
SYSTEMS

b.

Star Lighting Corp.

c.

ANDREWS
LABORATORIES

d.

VISION
technologies

e.

EASTERN
MEDICAL
associates

f.

Central Outfitters

THE DRAGLINE

Use additional paper as needed.

A dragline is a machine that works like a crane and is used to dig in ponds or lakes. The most important part of the dragline is the long arm with a hinge at the top like an elbow.

Shown below are two schematic drawings of a dragline.

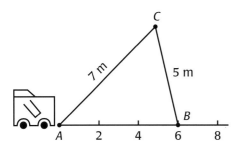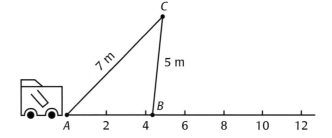

The dragline above has two arm sections, *AC* and *CB*. Point *A* is fixed to the machine, and *C* can move up or down. As *C* moves up or down, *B* moves either toward or away from the machine.

In the dragline above, arm section *AC* is 7 meters, and arm section *BC* is 5 meters.

1. Estimate the length of *AB* in the drawings above.

2. Use the scale above to draw a picture of when the arm sections of the machine reach out to 8 meters.

3. If you continue moving *B* to the right, the angle at *A* and the angle at *B* will become smaller. Will the angle at *A* always be smaller than the angle at *B*? Explain your answer.

4. a. Can the machine ever dig a hole 15 meters from *A*? Explain.

   b. Can the machine ever dig a hole 1 meter from *A*? Explain.

   c. Show in the drawing the maximum and minimum distances the arms can reach.

Section A.   Triangles Everywhere

1. Twenty triangles are part of the jungle gym's structure. The jungle gym has four sides, and each side is made up of five triangles (four small triangles and one large triangle). So, 4 × 5 = 20 triangles.

2. Drawings will vary. Sample drawings:

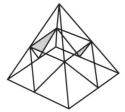 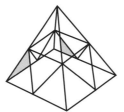

3.

4. Drawings may vary. Sample drawing:

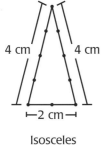

Section B.   The Sides

1. The following combinations of whole numbers add to 10:

1, 1, 8	2, 2, 6
1, 2, 7	2, 3, 5
1, 3, 6	2, 4, 4
1, 4, 5	3, 3, 4

2. a.

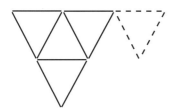

Isosceles   Isosceles

b. Only two of the possible combinations in problem **1** will form a triangle: 2, 4, 4 and 3, 3, 4. This is because, to form a triangle, the sum of the lengths of any two sides must be greater than the length of the third side.

Section C. The Angles

1.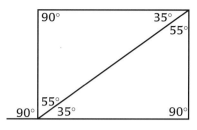

2. Drawings will vary. Sample drawing:

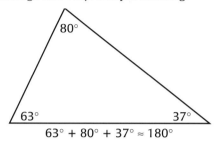

$63° + 80° + 37° ≈ 180°$

The angles were measured correctly if the sum of the angles equals 180°.

Section D. Sides and Angles

1. Drawings and triangles will vary. Sample drawing:

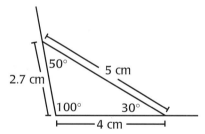

2. Yes, it is possible. Sample drawing:

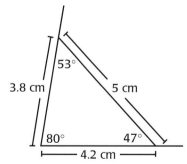

Students who answer, "No, it is not possible," may have tried to make one of the other sides too long, or one of the other angles larger than 80°.

3. It is not possible to draw a triangle in which the largest angle is 50° and the longest side is 5 centimeters. If one angle is 50°, then the other two angles must add to 130° because 180° − 50° = 130°. Since the two remaining angles must add to 130°, then at least one of the two remaining angles must be greater than 50° because 130° ÷ 2 = 65°. Therefore, the largest angle in a triangle can never be as small as 50°.

Section E. Parallel

1. a.

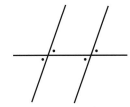

b. All the angles in the drawing that do not have a dot are equal to each other.

2. a. Drawings may vary. Sample drawing:

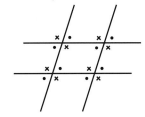

b. The shape shown above is a parallelogram because it is a four-sided figure formed by the intersection of two pairs of parallel lines. (Note: The shape is not a rhombus or a square because its sides are not equal in length. The shape is not a rectangle because it does not have right angles.)

3. Drawings may vary. Sample drawings are shown below.

a. A rectangle is a parallelogram with right angles.

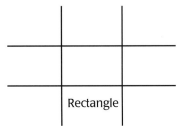

Rectangle

b. A square is a parallelogram with four right angles and four equal sides.

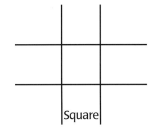

Square

c. A rhombus is a parallelogram with four equal sides.

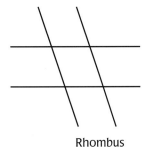

Rhombus

Section F. Copies

1. a. No. Joyce cannot fold her parallelogram in half so that the two parts fit together exactly because the parallelogram does not contain a line of reflection. So, it does not have line symmetry.

b. Yes. Any rectangle, rhombus, or square can be folded in half so that its two parts fit together exactly. They all contain a line of reflection and have line symmetry.

2. a. Drawings will vary. The sample drawing shown below illustrates a translation followed by a reflection:

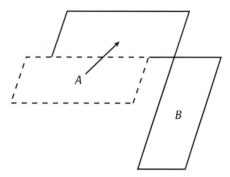

b. Yes. There are various solutions to part **a** because several movements of one type can have the same effect as one movement of another type. For example, several reflections may have the same result as rotating the original shape. Moreover, the sequence of movements can vary. For example, the above drawing could have been made by first using a reflection and then a translation.

c. A reflection is required for all possible solutions to part **a.**

Section G. Rotating Triangles

1. The letters H, I, N, O, S, X, and Z look the same when they are rotated 180°. (Note: Turn the page upside down to see which letters look the same.)

2. a. The vertex angle of the isosceles triangle measures 18° (360° ÷ 20 = 18°).

b. One inside angle of a 20-gon measures 162°. Since the three angles of a triangle total 180°, and since the vertex angle measures 18°, the remaining two angles total 180° − 18° = 162°. In an isosceles triangle, the remaining two angles are equal, and 162° divided by two is 81° per angle. Recall that the regular 20-gon was made by rotating the isosceles triangle. So, the inside angle of the 20-gon is made up of these two remaining angles, which is 81° + 81° = 162°. See the sample drawing shown below:

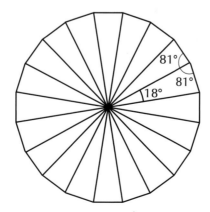

Cover

Design by Ralph Paquet/Encyclopædia Britannica Educational Corporation.

Collage by Koorosh Jamalpur/KJ Graphics.

Title Page

Phil Geib/Encyclopædia Britannica Educational Corporation.

Illustrations

6, 10 Phil Geib/Encyclopædia Britannica Educational Corporation; **14** Ralph Paquet/Encyclopædia Britannica Educational Corporation; **24, 26, 28, 30** Phil Geib/Encyclopædia Britannica Educational Corporation; **38, 42** Jerome Gordon; **52, 54, 68** Phil Geib/Encyclopædia Britannica Educational Corporation; **70** Jerome Gordon; **72** Phil Geib/Encyclopædia Britannica Educational Corporation; **76** Jerome Gordon; **84** Phil Geib/Encyclopædia Britannica Educational Corporation; **86** Paul Tucker/Encyclopædia Britannica Educational Corporation; **92** Phil Geib/Encyclopædia Britannica Educational Corporation; **98, 100** Jerome Gordon; **102** Paul Tucker/Encyclopædia Britannica Educational Corporation; **108, 110** Brent Cardillo/Encyclopædia Britannica Educational Corporation.

Photographs

8 © Jan de Lange; **32** © Ezz Westphal/Encyclopædia Britannica Educational Corporation; **66** © Larry Lefever/Grant Heilman Photography, Inc.; **74** © Doris DeWitt/Tony Stone Images.

Mathematics in Context is a registered trademark of Encyclopædia Britannica Educational Corporation. Other trademarks are registered trademarks of their respective owners.